I0048216

THE ROLE OF INTERMEDIARIES IN INCLUSIVE WATER AND SANITATION SERVICES FOR INFORMAL SETTLEMENTS IN ASIA AND THE PACIFIC

DECEMBER 2023

ADB

ASIAN DEVELOPMENT BANK

Notes:
In this publication, "$" refers to United States dollars.
ADB recognizes "Vietnam" as Viet Nam.

On the cover: **Inclusive essential services.** Intermediary service providers across Asia and the Pacific can
overcome water and sanitation delivery barriers to reach underserved populations (photos from top left to right by
Lester V. Ledesma/ADB, Amit Verma/ADB, Luis Enrique Ascui/ADB, Samir Jung Thapa/ADB, Abir Abdullah/ADB,
Lester V. Ledesma/ADB, and Eric Sales/ADB).

All photos are from the ADB Multimedia Library, unless otherwise noted.
Cover design by Ginojesu Pascua.

Printed on recycled paper

Contents

Tables, Figures, and Boxes

Filling the gap. Informal service providers, or intermediaries, have emerged in response to a significant essential service gap in informal settlements across Asia and the Pacific (photo by Eric Sales/ADB).

Acknowledgments

This publication has been prepared by the Strategy and Partnerships Team, Water and Urban Development Sector Office, Sectors Group, of the Asian Development Bank (ADB). It was led by **Neeta Pokhrel**, Director, Pacific and Southeast Asia Team (former chief of Water Sector Group); **Satoshi Ishii**, Director, Strategy and Partnerships Team; **Mafalda Pinto**, former Water Supply and Sanitation Specialist; and **Christian Walder**, former Water Supply and Sanitation Specialist. It was collated and authored by **Isabel Blackett** (ADB Consultant) and **Penny Dutton** (ADB Consultant) with extensive contributions on core data, reports, and information for the case studies from the following organizations: Eau et Vie, Toilet Board Coalition, 1001fontaines, and WaterAid. In addition, many ADB staff have contributed to this publication, including ADB's Water Advisory Team on Sanitation. Special thanks given to **Saswati Belliappa** (ADB Senior Safeguards Specialist, Office of Safeguards) for peer reviewing this publication and to **Joanna Brewster** (ADB Consultant) for exemplary editing. Other ADB staff and consultants who contributed to this publication include **Vikkie Antonio** and **Joy Bailey** (ADB Urban Climate Change Resilience Trust Fund Consultants); **Lyn Almario** (former ADB Consultant, Water Sector Group); **Stephane Bessadi** (ADB Senior Procurement Specialist, Procurement, Portfolio, and Financial Management Department); and **Alain Morel** (ADB Principal Country Specialist, Southeast Asia Department). Colleagues from ADB's Human and Social Development Sector Office also provided valuable suggestions and guidance during its development.

ADB's Strategy and Partnerships Team is grateful to the following organizations working on the ground that inspired this publication: Eau et Vie (Bangladesh and the Philippines), Tubig Pag-Asa, Inc. (Philippines), Shobar Jonno Pani (Bangladesh), Bhumijo (Bangladesh), Suvidha (India), Aerosan (Nepal), Laguna Water (Philippines), 1001fontaines (Cambodia), and Motu Koita Assembly (Papua New Guinea).

Abbreviations

ADB	Asian Development Bank
CBO	community-based organization
COVID-19	coronavirus disease
DEWATS	decentralized wastewater treatment system
DFI	development finance institution
FGD	focus group discussion
FSM	fecal sludge management
IT	information technology
IWADCO	Inpart Waterworks and Development Company
MCWD	Metro Cebu Water District
MKA	Motu Koita Assembly
MOA	memorandum of agreement
NGO	nongovernment organization
OECD	Organisation for Economic Co-operation and Development
PNG	Papua New Guinea
QR	quick response (code)
SANIMAS	Sanitasi Berbasis Masyarakat (Sanitation by Community)
SDG	Sustainable Development Goal
SJP	Shobar Jonno Pani
SMEs	small and medium-sized enterprises
TBC	Toilet Board Coalition
TPA	Tubig Pag-Asa, Inc.
UN-Habitat	United Nations Human Settlements Programme
WAF	Water Authority of Fiji

Executive Summary

The Asia and Pacific region is home to more than half of the world's urban informal settlement population—about 666 million people. Although declining as a proportion of the urban population, the number of people living in settlements is growing.

To achieve the Sustainable Development Goals, all urban residents need access to safe and sustainable water and sanitation systems. Yet many people in these settlements lack access to a safe water supply and to basic or even limited levels of sanitation. Providing water supply and sanitation services to these settlements is challenging because of a lack of interest by utilities, marginal land, high-density living, and low-income levels to pay for services. Usually, service needs are met by informal service providers.

The Asian Development Bank (ADB) document, *Strategy 2030 Water Sector Directional Guide: A Water-Secure and Resilient Asia and the Pacific* envisions no one is left behind in ensuring basic services, including water supply and sanitation. In line with the priorities and goals of these strategies, this publication examines good practices in the Asia and Pacific region, in particular, those involving the role of intermediaries—social enterprises, the private sector, and nongovernment organizations—in delivering water and sanitation services in informal settlements and their potential for expansion. It presents eight case studies from Bangladesh, Cambodia, India, Nepal, Papua New Guinea, and the Philippines as examples of using intermediaries.

This publication explores informal settlements' characteristics and the demand and supply of water and sanitation services. Recognizing that such settlements vary greatly within a city, between cities in the same country, and between countries, this publication uses new and previously undocumented research to demonstrate that working with different types of intermediaries provides a viable option for formal utilities and local governments to provide services to underserved or excluded residents. In return, the newly served areas can help grow the customer base, increase revenue, meet service targets, and, in some situations, reduce nonrevenue water.

The new case studies reveal that intermediaries share common features in how they operate and approach service delivery, including formal ties with local governments and utilities, support from parent organizations, deep community engagement and presence, built-in capacity development, and user pays services that ensure financial viability for operations while being sensitive to the needs of the lowest-income households.

All intermediaries in the case studies deliver high-quality services while adapting and innovating their services to use technology for lower cost, service efficiency, and resilience. Intermediaries demonstrated that they can quickly adapt to difficulties such as the coronavirus disease (COVID-19) pandemic and long-term threats such as climate change. All case studies show potential to scale, with some already operating across multiple locations and serving many thousands of customers.

This publication provides new guidance to development finance institutions (DFIs) on where and when an intermediary could be suitable within a large urban water or sanitation investment project. It also provides advice on desirable qualities in intermediaries—such as being locally based, technically competent, adaptable, and motivated by social objectives and outcomes. Issues related to contracting intermediaries and a suggested timeline (4–5 years) are outlined. Risks and challenges are realistically presented with suggestions to mitigate these. Recommendations are made on how DFIs can support service delivery to informal settlements by adapting their approaches and systems.

The lessons from the case studies are intended to encourage governments, water supply and sanitation utilities, and development financing institutions to incorporate intermediate service providers as valuable partners in large-scale investments in water and sanitation service delivery.

This publication targets staff of DFIs, consulting firms, and governments seeking models and options to improve water and sanitation services to informal urban settlements in the context of wider urban water and sanitation investments. It aims to support the design of water and sanitation investment projects, which include providing viable and sustainable services to informal settlements. Without serving all, including those in the informal settlements, the development objective in ADB's Strategy 2030 and Water Sector Directional Guide of leaving no one behind in ensuring basic services will not be met.

1

Introduction

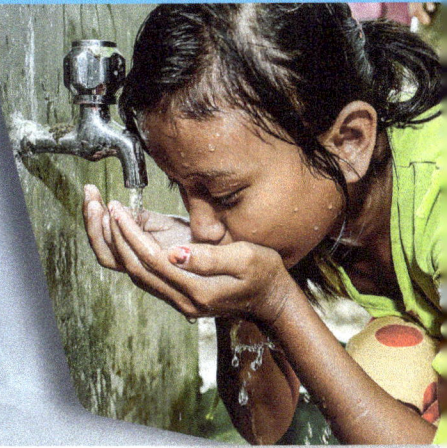

The Urban Water and Sanitation Challenge

More than half of the world's urban informal settlement population—about 666 million people (63%)—live in Asia and the Pacific (United Nations Human Settlements Programme [UN-Habitat]).[1] While the number of people living in settlements, as a proportion of urban inhabitants, is diminishing, the number of people living in settlements is growing. As these settlements grow, so does the need for safe and sustainable water and sanitation services. Yet, about 500 million people in Asia and the Pacific lack access to a safe water supply, and 1.14 billion people lack access to basic or even limited levels of sanitation (Asian Development Bank [ADB] 2022).

Accessing clean water and sanitation is a constant challenge for urban residents in informal settlements and marginal areas (United Nations General Assembly 2010).[2] For most residents, informal local water and sanitation service providers meet these basic needs. These local entrepreneurs or enterprises provide services paid for directly by the clients. They are not planned, authorized, supervised, or acknowledged by the formal authorities as part of the official system (IGI Global 2021). These services include water tankers, bottled water delivery, provision and management of shared or community latrines, unregistered pit emptying, container-based sanitation, or piped water to a private household or shared tap. These informal service providers, or intermediaries, have emerged in response to a significant essential service gap (Garrick et al. 2019).[3]

How intermediaries operate between formal providers, such as utilities and consumers, is largely hidden in mainstream service delivery models. This publication highlights these service providers by sharing case studies and examples of extending water and sanitation services to South Asia and Southeast Asia and Pacific informal settlements. Case studies from Bangladesh, Cambodia, India, Nepal, Papua New Guinea, and the Philippines and experience from the contributors to this publication show that intermediary service providers frequently overcome the water and sanitation service delivery barriers formal providers face. By doing so, they deliver services in informal settlements through innovative, flexible, and community-responsive models, including during the challenges of the coronavirus disease (COVID-19) pandemic.

---

[1] In general, the term "informal settlements" is used in this publication to encompass all definitions, including slums and squatter settlements. Local terms for informal settlements are used for some contexts.

[2] Informal settlements are unplanned areas that lack secure tenure, basic services, and urban infrastructure. Marginal areas include settlements on steep slopes, floodplains, riverbanks, and mangroves.

[3] The term "intermediaries" is used in different ways (for example, see Lardoux de Pazzis and Muret 2021). In this publication, it means small and medium-sized enterprises (SMEs) comprising private companies, social enterprises, nongovernment organizations (NGOs), and community-based organizations (CBOs). The term is also used to describe the function intermediaries play between a formal utility or government provider and communities. It is not biased toward nor does it imply a particular legal structure or arrangement.

Navigating This Publication

This publication aims to support the design of water and sanitation investment projects, which include providing viable and sustainable services to informal settlements. It aligns with development objectives in ADB's Strategy 2030 and Water Sector Directional Guide, which envision no one left behind in ensuring basic services—including water supply and sanitation.[4]

This publication targets the staff of development financing institutions (DFIs), consulting firms, and governments seeking models and options to improve water and sanitation services to informal urban settlements in the context of wider urban water and sanitation investments.

It is assumed that staff and consultants using this publication are aware of some of the challenges, constraints, institutions, and context of the towns and cities where they plan to improve services. While context is critical, this publication focuses on practical examples and service delivery features that can be applied to a well-considered context.

This publication outlines in section 2 the demand and supply for informal water and sanitation services in Asia and the Pacific (Figure 1, level 1). The supply side (section 2.3) focuses on providing services by small-scale providers in South Asia, Southeast Asia, and the Pacific (Figure 1, level 2). Section 3 describes common approaches seen in improving informal water and sanitation services.

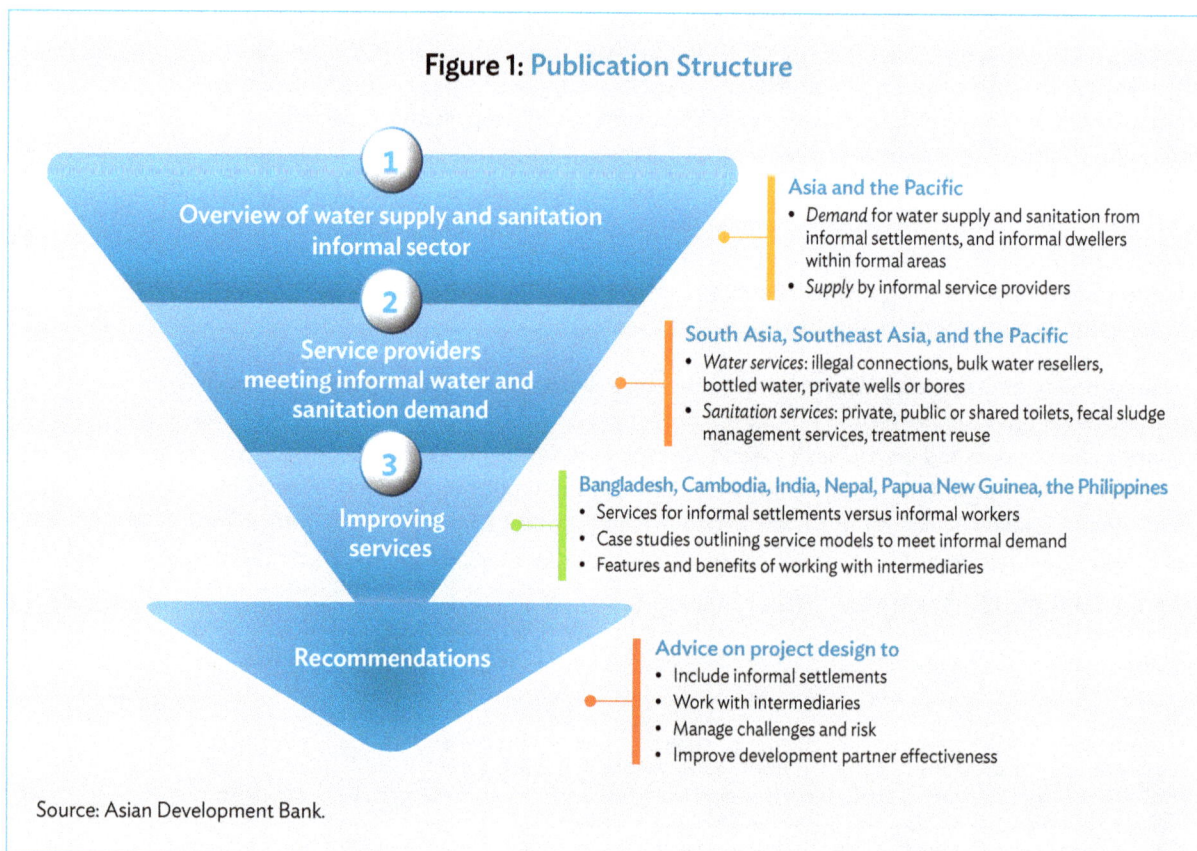

Figure 1: Publication Structure

1 Overview of water supply and sanitation informal sector

Asia and the Pacific
- *Demand* for water supply and sanitation from informal settlements, and informal dwellers within formal areas
- *Supply* by informal service providers

2 Service providers meeting informal water and sanitation demand

South Asia, Southeast Asia, and the Pacific
- *Water services*: illegal connections, bulk water resellers, bottled water, private wells or bores
- *Sanitation services*: private, public or shared toilets, fecal sludge management services, treatment reuse

3 Improving services

Bangladesh, Cambodia, India, Nepal, Papua New Guinea, the Philippines
- Services for informal settlements versus informal workers
- Case studies outlining service models to meet informal demand
- Features and benefits of working with intermediaries

Recommendations

Advice on project design to
- Include informal settlements
- Work with intermediaries
- Manage challenges and risk
- Improve development partner effectiveness

Source: Asian Development Bank.

4   ADB. 2018. *Strategy 2030: Achieving a Prosperous, Inclusive, Resilient, and Sustainable Asia and the Pacific*. Manila; and ADB. 2022. *Strategy 2030 Water Sector Directional Guide—A Water-Secure and Resilient Asia and the Pacific*. Manila.

This background provides the context in section 4 for presenting the case studies of water and sanitation services in informal communities in six countries (Figure 1, level 3). These include responses to the COVID-19 pandemic and how digital technology has supported this response.

Section 5 brings together the common elements from the case studies. Practical recommendations are made for how ADB and other DFIs can serve informal settlements with reliable, sustainable, and quality water and sanitation services within investment projects.

Research Approach and Methods

This publication primarily draws on eight case studies from six countries (Bangladesh, Cambodia, India, Nepal, Papua New Guinea, and the Philippines) to illuminate examples of how informal water and sanitation demand is being met. Data for six case studies are drawn from ADB-commissioned fieldwork conducted by the Toilet Board Coalition (TBC) and Eau et Vie from October to November 2021.[5] An additional case study on Southeast Asia examines a rural water kiosk model in Cambodia, with information from reports by 1001fontaines.[6] The final case study focuses on the Pacific by highlighting a water supply demonstration village in Papua New Guinea.

The primary data were collected through surveys, semi-structured key informant interviews, and focus group discussions (FGDs). Gender inclusion was an important element in the survey and FGD methodology, with data disaggregated by sex where possible. Information on COVID-19 pandemic impacts and responses were also collected. In addition, secondary data sources were used, such as operational reports, and other publications.

The case studies and data collection methods, along with a description of the services, the users, and the relationship with formal services, are summarized in Table 1. A full description of the case studies is provided in Appendix 1. Additional notes on the methodology are presented in Appendix 2.

The case studies are an important, but not the only, source of information. The models proposed have also drawn on published and unpublished data sources, including other case studies, interviews, and learning from research, practice, and experience of ADB staff and consultants, who contributed to this publication.[7]

There are limitations to the information in this publication. The case studies are specific to their context and, thus, may not apply broadly to other situations. The primary research was also impacted by the COVID-19 pandemic, which required fieldwork to be conducted using social distancing and virtual communications, following local regulations effective at the time. In addition, the case studies present a practical perspective and examples of delivering water and sanitation services in informal settlements and, thus, do not provide comprehensive documentation of potential models to consider.

---

[5]   TBC is a business-led, membership-based, global nonprofit organization that works with private businesses, entrepreneurs, investors, and governments to scale up market-based solutions to progress toward universal access to sanitation. Eau et Vie is a French NGO that helps establish social enterprises and local associations to improve water, sanitation, and waste conditions in deprived neighborhoods of Bangladesh and the Philippines.

[6]   1001fontaines is a French NGO working to improve the health of vulnerable populations by supporting small enterprises to produce and deliver affordable, safe drinking water.

[7]   See references section for full list of sources.

Table 1: Summary of Water and Sanitation Case Studies

Service Provider	Location	Summary of Service(s) Provided	Service Area and Population Served	Relationship to Formal Services	Data Collection		
					Survey	KII	FGD
TPA Social business	Guizo, Mandaue City, Cebu, Philippines	Household piped water connections	1,400 people/308 households served by 282 connections	TPA buys bulk water from MCWD under a 10-year memorandum of agreement		x	x
SJP Social enterprise	9 Number Bridge, Chattogram, Bangladesh	Household piped water connections, household latrines, or shared toilets blocks	1,900 people/447 households served by 450 contracts	SJP buys bulk water from CWSA		x	x
Bhumijo Social enterprise	Dhaka, Rangpur, Khulna, Narayanganj, Bangladesh	Community and public toilet operations	250–260 users/day More than 15 facilities at more than 6,000 people/day	Connected to sewerage if available; land provided by local governments	x	x	x
Suvidha Partnership of Hindustan Unilever with MCGM	Mumbai, India	Community toilet, laundry services, and water kiosk	Seven centers serving a total of 200,000 people per year (550 people/day)	Partnership with the MCGM and HSBC India	x	x	x
Aerosan Social enterprise	Swoyambhu, Nepal	Female-focused and inclusive community sanitation facility with treatment	Six locations with approximately 400 users/day pre-COVID-19 pandemic	Working with local government on the provision of land during the COVID-19 pandemic	x	x	x

continued on next page

Table 1 continued

Service Provider	Location	Summary of Service(s) Provided	Service Area and Population Served	Relationship to Formal Services	Data Collection		
					Survey	KII	FGD
LAGUNA WATER Laguna Water Joint venture of PGL and Manila Water	Barangay San Antonio, Santa Rosa City, Laguna, Philippines	Sanitation and water service utility, including desludging of septic tanks	Desludging services provided to Laguna Water customers; 5,500 households	Laguna Water is a formal utility	x	x	x
1001FONTAINES 1001fontaines Social enterprise working with local entrepreneurs	Cambodia	Water kiosks with a 20-liter container drinking water delivery service	270 water services supplying 25% of the rural population	Extracts and treats water in cooperation with local governments	Report only		
Motu Koita Assembly Social business	Port Moresby, Papua New Guinea	Piped water to kiosks for container sales and future household connections	550 households	MKA buys bulk water from WPNG	Report only		

COVID-19 = coronavirus disease, CWSA = Chattogram Water and Sewerage Authority, FGD = focus group discussion, KII = key informant interview, MCGM = Municipal Corporation of Greater Mumbai, MCWD = Metropolitan Cebu Water District, MKA = Motu Koita Assembly, PGL = Provincial Government of Laguna, SJP = Shobar Jonno Pani, TPA = Tubig Pag-Asa, Inc., WPNG = Water PNG.

Sources: Authors and Asian Development Bank.

Opportunity for action. More than half of the world's urban informal settlement population—about 666 million people (63%)—live in Asia and the Pacific and face constant challenges in accessing clean water and sanitation (photo by Penny Dutton/WaterAid).

2

Overview of the Informal Sector

Definitions and Characteristics of Settlements and Services

The following are descriptions of some key terms used in this document.

Informal settlements and slums. Informal settlements are residential areas where inhabitants lack security of tenure, basic services, and urban infrastructure. Their dwellings are unlikely to comply with planning and building regulations and are often situated in geographically and environmentally challenging or hazardous areas. Slums are the most deprived and excluded informal settlements, characterized by poverty and agglomerations of dilapidated and semi-permanent housing, often located in the least desirable urban areas, with people living in slums exposed to eviction, disease, and violence (UN-Habitat 2015).

Formal settlements and housing. Formal settlements are zones or areas designated for residential housing, usually in an urban plan. The land is legally registered and formally owned and typically has ready access to basic urban infrastructure and services, such as roads, water and sewerage, electricity, refuse collection, and street lighting. Housing is usually built to meet planning regulations and building codes.

Informal water and sanitation service providers. Informal water and sanitation service providers are individual entrepreneurs or enterprises that provide a water or sanitation service paid for by the clients, but not planned, authorized, supervised, paid, or acknowledged by the formal authorities as part of the official system (IGI Global 2021). Examples of informal service providers include water vendors (by bottles, containers, carts, or tankers), private borehole or well water sellers, unregistered latrine builders and plumbers, retailers of non-standardized products, providers of illegal connections to water or sewerage mains, and manual or mechanical emptiers of pit latrines and septic tanks.

Formal water and sanitation service providers. Formal service providers are registered, legal companies that usually pay taxes. The term primarily refers to government-owned utilities and businesses that operate under regulation and provide mostly piped services. However, it can also include private sector businesses formally contracted, delegated, regulated, and/or licensed to provide the services.

Typical characteristics of informal and formal settlements and water supply and sanitation services are summarized in Table 2.

Table 2: Typical Characteristics of Informal and Formal Settlements and Services

	Informal	Formal
Settlements and housing	• Unplanned, unzoned • Limited or no government services • Basic housing often built with non-permanent materials • High population density and overcrowding, or lower density in peri-urban areas • Often on marginal land, e.g., flood-prone, steep slopes, or over water bodies • Mixed income levels (usually low or very low income) of residents	• Cadastral surveyed land • Planned settlements with laid-out, defined housing plots • Registered owners • Formal services: roads, water, sewerage, power, transport, and telecommunications • Low-to-medium population density • Mixed income levels (usually middle and higher income) of residents
Water and sanitation services	• Self-provision or service by informal service providers • Shared or communal facilities, e.g., toilets and water points • Overused services, sometimes requiring queuing • Unregistered or unknown water sources and poor-quality water • Treated or untreated water supplies • Low-quality pits or tanks • Unreliable services of variable quality In some places, there are also: • Illegal connections • Extortionate control of services and prices	• Formal provision by government entities and utilities or registered, designated private service providers • Household-level services to house or plot • Piped, treated water • Licensed service providers • Maintained and reliable services • Controlled prices through regulation • Accountability (of utilities to regulators)

Source: Asian Development Bank.

Informal settlements vary enormously. Their specific characteristics depend on the local and geographic context. Many variables and differences exist between countries, between cities, and even between informal settlements in the same city. In addition, evidence indicates it is inaccurate to stereotype informal services as illegal, inefficient, traditional, expensive, or unsustainable. They often co-produce with formal services and may include stable, dynamic, expanding, and profitable businesses (Ahlers et al. 2014). While there are some typical characteristics of informal settlements and services, as the case studies show, the reality is often more complex, and the boundaries of informal and formal are blurred.

Understanding the informal demand for services and the existing informal service providers in the proposed service area is necessary to design appropriate, affordable, and sustainable improvements in access to water and sanitation services.

Understanding Informal Demand

Informal demand for water supply and sanitation typically comes from informal areas such as squatter or slum settlements and informal businesses. In a city's formal areas, resident and mobile populations may also demand informal water and sanitation services (Table 3).

Table 3: Sources of Informal Demand for Water and Sanitation Services

Informal Demand for Services	
From Informal Areas	**From Formal Areas**
• Residents of extensive informal settlements • Workers in informal businesses	• Residents of formal settlements without reliable water and sanitation services • Residents of clusters of informal housing within formal areas • Homeless people in city centers and other formal areas • Informal workers and street vendors working in formal areas

Source: Asian Development Bank.

Informal Demand in Informal Areas

Informal demand comes from households that cannot access the formal water supply and sanitation services provided by water and sewerage or sanitation utilities or the registered, legal private sector. Reasons for lack of access may include the legal, physical, and perceived barriers of the water and sanitation utility, household living circumstances, low income, and unaffordability of services.

Typical barriers of utilities to providing services and of users to accessing services are identified in Table 4.

The most sizable informal demand is from the estimated 666 million residents of informal settlements across Asia and Pacific region (UN-Habitat). These settlements lack adequate, reliable, safe, and affordable water and sanitation services, and most residents rely on shared facilities such as common toilets or standpipes (Sinharoy, Pittluck, and Clasen 2019).

Table 4: Common Utility and User Barriers to Services

Utility Barriers to Providing Water Supply and Sanitation Services in Informal Communities	
• Water and sanitation service networks are limited and do not include the whole city or town • Lack of tenure (ownership) or landlord agreement • Utility mandate excludes informal areas and does not include the full water or sanitation service chain • Fear of legitimizing settlements by providing services • Inadequate bulk water (or treatment capacity) to extend access to formal or informal communities • High-density unplanned layouts are difficult for infrastructure and access (e.g., pipelines and roads for fecal sludge management trucks) • Fecal sludge management treatment facilities are unavailable near the service area • Limited capital resources to expand services into new areas	• Underdeveloped billing and revenue collection systems that limit operating revenue • A billing address is required • Political lack of will to charge adequate water and sanitation tariffs • Perception that customers living below the poverty line cannot pay • High service levels are expensive and do not meet the demand • Concern about increasing unrecoverable operation and maintenance costs • Lack of experience working in low-income communities • Violence, community conflict, or personal safety risks for utility staff

continued on next page

Table 4 *continued*

User Barriers to Accessing Water Supply and Sanitation Services in Informal Communities	
Water services	**Sanitation services**
• Lack of piped water service in the area • Housing tenant (renter), lack of house or land tenure, or the landlord does not permit connections to piped water services • Multistory dwelling with no piped water service • Unaffordable water connection charge • Unaffordable water tariff • Lack of address for regular bills	• Lack of sewers in the area • Lack of space to build a latrine • Housing tenant (renter), lack of house or land tenure, or the landlord does not permit connections to sewerage • Multistory dwelling with no sewers • Unaffordable sewer connection charge • Unaffordable sewerage tariff or desludging charge • No access road in dense housing areas to empty pits or tanks • High groundwater

Source: Conan, H. 2004. Small Piped Water Networks: Helping Entrepreneurs to Invest. *ADB Water for Life Series.* No. 13. Manila: Asian Development Bank.

Informal Demand in Formal Areas

Informal demand from within formal settlements occurs when services are unreliable or inadequate. If sewers and licensed fecal sludge management (FSM) services are unavailable to formal housing and businesses, demand for informal services is created. Similarly, if the utility piped water supply is intermittent or unreliable, significant demand for other water services, such as water delivered by cart or tanker and household storage, is created.

Formal areas may also include demand from people who are homeless or squatting in city centers (McIntosh 2003).

In some Asian countries, for example, Singapore and Malaysia, with high levels of quality, affordable, and reliable urban water and sanitation access, informal demand is low. However, in most countries— including these and Organisation for Economic Co-operation and Development (OECD) countries— informal demand is high among homeless people, migrants, and unregistered occupants of illegal or semi-legal properties. Official statistics often exclude homeless people and migrants.

Some cities have small informal settlements among formal middle-class areas, often comprising domestic workers, gardeners, guards, and other service workers. Many informal workers operate in the city center and middle-class areas or neighboring informal areas. Informal workers require access to water and sanitation services while working and at the terminuses of their commutes. In some cities, tens or hundreds of thousands of day workers commute long distances from rural villages to trade or work low-paying jobs in formal areas.

Demand may include the needs of food and drink street vendors and market traders, as well as service providers, including those working in transport, refuse collection, waste picking, and recycling, who often work far from their homes. Figure 2 outlines the needs of informal workers.

Figure 2: What Informal Workers Need from Inclusive Cities

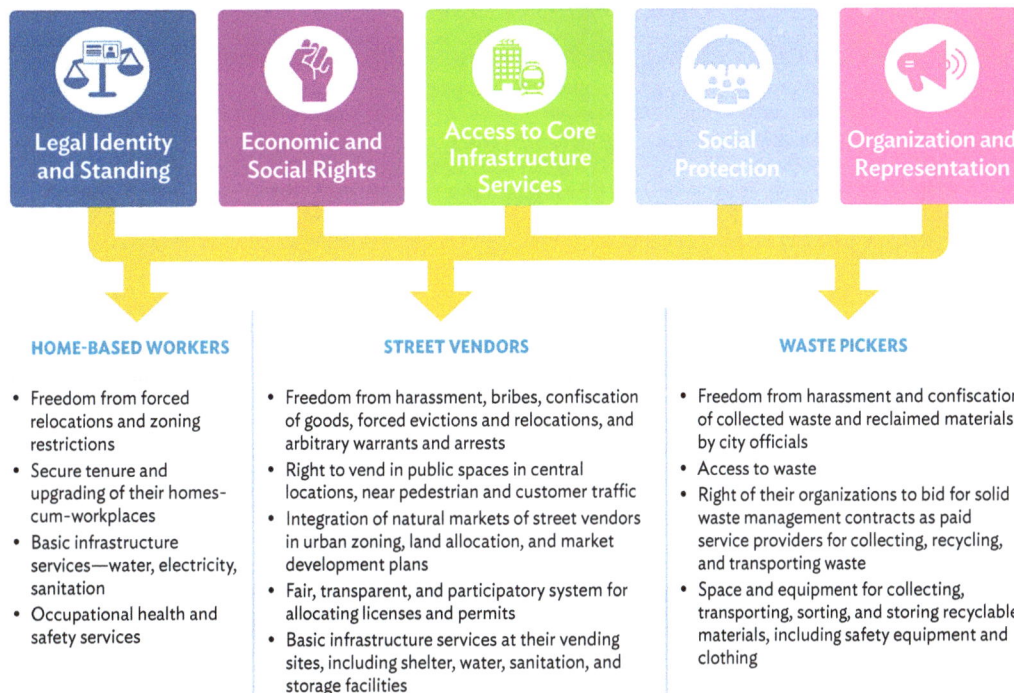

Legal Identity and Standing	Economic and Social Rights	Access to Core Infrastructure Services	Social Protection	Organization and Representation

HOME-BASED WORKERS

- Freedom from forced relocations and zoning restrictions
- Secure tenure and upgrading of their homes-cum-workplaces
- Basic infrastructure services—water, electricity, sanitation
- Occupational health and safety services

STREET VENDORS

- Freedom from harassment, bribes, confiscation of goods, forced evictions and relocations, and arbitrary warrants and arrests
- Right to vend in public spaces in central locations, near pedestrian and customer traffic
- Integration of natural markets of street vendors in urban zoning, land allocation, and market development plans
- Fair, transparent, and participatory system for allocating licenses and permits
- Basic infrastructure services at their vending sites, including shelter, water, sanitation, and storage facilities

WASTE PICKERS

- Freedom from harassment and confiscation of collected waste and reclaimed materials by city officials
- Access to waste
- Right of their organizations to bid for solid waste management contracts as paid service providers for collecting, recycling, and transporting waste
- Space and equipment for collecting, transporting, sorting, and storing recyclable materials, including safety equipment and clothing

Source: Chen, M.A. and V.A. Beard. 2018. *Including the Excluded: Supporting Informal Workers for More Equal and Productive Cities in the Global South.* Working Paper. Washington, DC: World Resources Institute.

Informal Water and Sanitation Services

Providers of informal water and sanitation services across Asia and the Pacific often fill formal sector service provision gaps. The need, extent, and nature of the supply of informal services vary across countries and within cities. Factors influencing informal services include the local access, quality, reliability, and affordability of formal piped (or non-piped) water and sanitation services.

Informal water and sanitation markets vary greatly. They span simple and complex services involving diverse informal providers using different operational and service delivery models. Informal services are supplied throughout the water and sanitation service chains.

Sanitation

Informal sanitation services exist across the sanitation service chain (Figure 3). Without formal sewerage systems or regulated FSM services, residents and businesses have no choice but to self-supply to avoid open defecation.

The self-supply of sanitation includes residents and businesses constructing substandard latrines, pits, and septic tanks and paying for septage removal by informal manual or mechanical FSM service providers. Such informal sanitation services are ubiquitous in informal settlements. They are also

Figure 3: Typical Informal Services along the Sanitation Service Chain

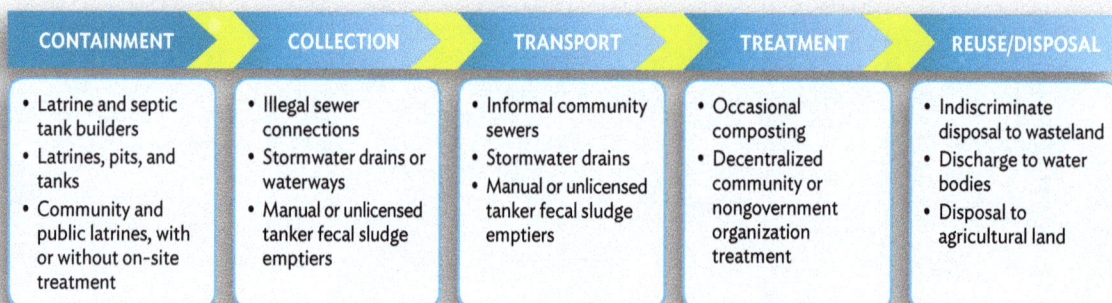

CONTAINMENT	COLLECTION	TRANSPORT	TREATMENT	REUSE/DISPOSAL
• Latrine and septic tank builders • Latrines, pits, and tanks • Community and public latrines, with or without on-site treatment	• Illegal sewer connections • Stormwater drains or waterways • Manual or unlicensed tanker fecal sludge emptiers	• Informal community sewers • Stormwater drains • Manual or unlicensed tanker fecal sludge emptiers	• Occasional composting • Decentralized community or nongovernment organization treatment	• Indiscriminate disposal to wasteland • Discharge to water bodies • Disposal to agricultural land

Source: Asian Development Bank.

used in many established city centers and formal residential areas that lack sewerage. Self-supply also includes wastewater disposal through illegal sewer connections or direct overflow (from toilets, septic tanks, or pits) to stormwater drains and water bodies. These practices are common in the megacities of Dhaka, Jakarta, and Manila.

Informal household sanitation examples include unhygienic hanging latrines in Manila, Philippines; Pekalongan and Tegal, Central Java, Indonesia; and Dhaka, Bangladesh and unimproved, unhygienic, dry pit latrines in India's towns and Ambon, Indonesia.[8]

Providers of informal FSM services include people from low castes or those living below the poverty line who manually empty latrines

Informal sanitation. Overhung or hanging latrines, such as this one in Pekalongan, Indonesia, are suspended above water bodies; they are unhygienic and pollute rivers, canals, and the sea (photo by Isabel Blackett).

in Bangladesh and India (World Bank et al. 2019) and unregistered companies that empty latrines and septic tanks with vacuum trucks in Indonesia and the Philippines (Shagun 2019).

There is also a market for untreated fecal sludge—as a valuable agricultural soil supplement—from urban areas. Such unsafe disposal is practiced on a small scale in many Asian countries (AECOM International Development, Inc. and Eawag, 2010).

Formal–Informal Sanitation Linkages

Informal sanitation services can be connected to formal services across the sanitation chain.

Common places where formal and informal sanitation services interact include the following:

---

[8]   A hanging toilet or latrine is a toilet built over a sea, river, or other body of water into which excreta drops directly (UNICEF and WHO 2018).

(i) **Illegal connections to sewerage and drainage systems.** Households and businesses informally or illegally connect their wastewater outlets to the storm drain or sewer systems, for example, in Siem Reap, Cambodia (Cambodia News English 2021).

(ii) **Emptying septic tanks or pits.** Septic tanks or pits are not built to standards or are constructed by informal builders or households; authorities do not inspect construction, or no formal system exists for registering on-site sanitation systems. These substandard, unregistered septic tanks and pits are emptied by registered FSM service providers. Such challenges are common in many Asian cities, including Banjarmasin, Jakarta, and Bandung in Indonesia; Thu Dau Mot and Da Lat in Viet Nam (World Bank Group and Water Sanitation Program 2015); Siem Reap in Cambodia; and Dhaka in Bangladesh.

(iii) **Disposal of waste.** Informal FSM service providers discharge their vacuum trucks of fecal waste into formal treatment facilities, such as utility sewage or fecal sludge treatment works.

(iv) **Informal sewage discharge.** Well-built toilets and septic tanks discharge informally into stormwater drains and water bodies because of a lack of space for on-site disposal, the absence of sewers, or expensive or obscure processes for connecting legally to a sewer. Such connections are seen in Surakarta (Solo), Indonesia; Phnom Penh, Cambodia; and Dhaka, Bangladesh (DOHWA Engineering Co. Ltd. 2021).

(v) **Informal tanker services.** Septic tanks are emptied by informal manual or vacuum tanker services, including, for example, informal pit emptiers from low castes in cities in India and Bangladesh (World Bank et al. 2019).

(vi) **Fecal sludge dumping.** Licensed FSM service providers dump fecal sludge indiscriminately onto land, water bodies, or agricultural land. These practices may be because no fecal sludge treatment is available, the treatment plant is located too far (more than a 30-minute haulage distance) or its limited hours of operation are incompatible with the services provided (for example, evening or early morning), or disposal is too expensive. In some countries, such as Cambodia, the Lao People's Democratic Republic (Lao PDR), and Nepal, farmers pay tanker drivers to dispose of fecal sludge on their land.

Water Services

Informal water supply services are also seen along the entire supply chain, often connected to the services provided by water utilities (Figure 4). Informal small-scale providers fill gaps left by inadequate piped water supplies and may constitute a significant market share in some cities. For example, Karachi, Pakistan has piped water available only 3 hours a day and not every day, resulting in 20% of households relying on informal tanker delivery and about 30% using other informal supply methods (Mitlin et al. 2019). Moreover, informal services may be the only option in fragile and conflict-affected areas (McIntosh 2003). In middle-upper-class neighborhoods, water may be informally sold from a formal connection or from utility-supplied water stored in a household (McIntosh 2003). In some cities, such as Shanghai,

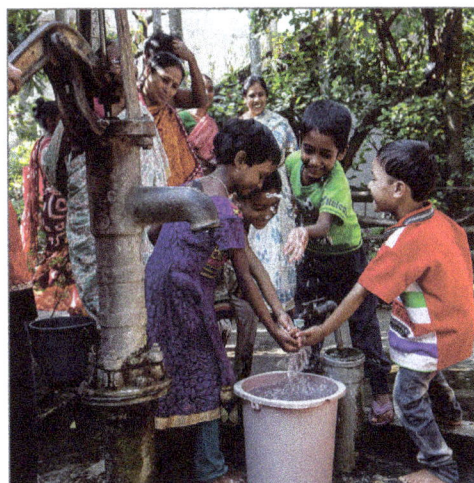

Informal water supply. For urban settlement residents in India, water service delivery can involve small and medium-sized enterprises as intermediaries between them and the formal sector (photo by Amit Verma/ADB).

Figure 4: Typical Informal Services along the Water Supply Service Chain

EXTRACTION	TREATMENT	STORAGE	DISTRIBUTION	RETAIL
• Community borehole or well • Private bores or wells • Tanker services • Utility bulk water	• Small piped network dosing • Bottled water manufacturers • Household treatment	• Private tanks (e.g., rainwater, underground) • Informal piped network tanks	• Bicycle vendors • Pushcart operators • Motorbike vendors • Tanker truck service • Private small pipeline	• Illegal connections to utility • House connections • Standpipe • Kiosk • Bottled or sachet water

Source: Asian Development Bank.

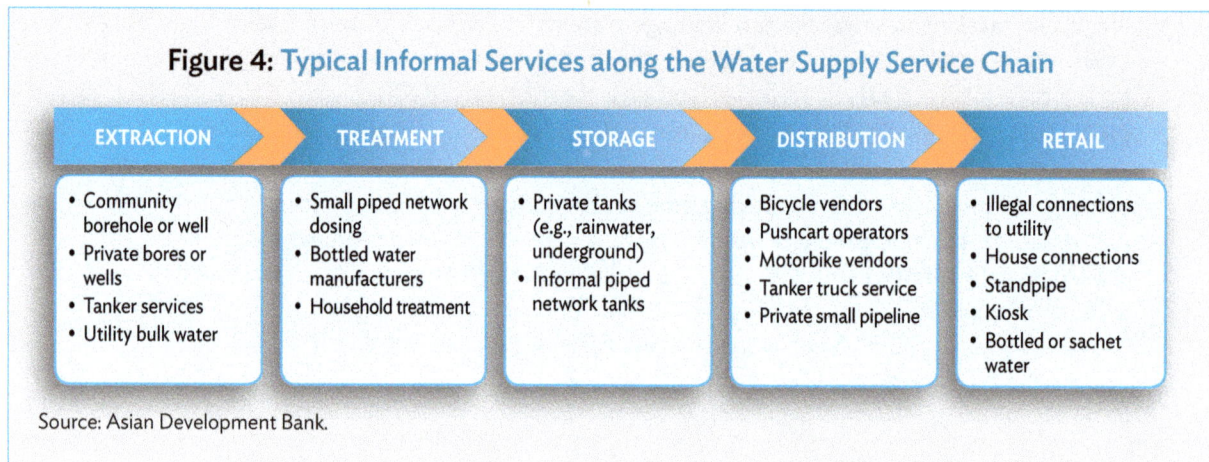

where the utility provides residents with reliable and affordable water service, informal street vendors may be limited to bottled water sales (McIntosh 2003).

Informal water providers in Asia and the Pacific's urban areas are highly heterogeneous. Services include private companies or cooperatives serving thousands of households through small, piped networks or tanker truck delivery (for instance, in Delhi and Ho Chi Minh City), community-initiated neighborhood providers of basic water services (paid or free), resellers (in Manila, Cebu, and Jakarta), street vendors of untreated water such as in South Asia, and bottled water sellers in tourist hot spots such as Siem Reap, Cambodia and Bali, Indonesia. Most providers are independent and entrepreneurial, without formal recognition from local authorities, yet effective at serving hard-to-reach places in urban areas (McIntosh 2003).

A smaller and overlooked group of informal water providers are self-supplied households and businesses. Self-suppliers harness hand-dug wells, tube wells, boreholes, and rainwater to meet their water needs. They may use different water sources and qualities for different purposes, such as well water for washing clothes and piped or bottled water for drinking and cooking. Examples are found in major Asian cities and in medium-sized towns in Indonesia, the Lao PDR, and Viet Nam.

Formal–Informal Water Supply Linkages

For water supply, the linkages between the formal and informal sectors are usually in one direction, through water supplied from a formal utility (or business) to informal service providers. These linkages where formal and informal water supply services interact include

(i)   connecting an informal piped community network to a bulk water supply from a utility;

(ii)   illegal offtakes from a utility pipeline by water vendors;

(iii) regulation of water sources, for example, licensing a local community borehole; and

(iv)   informal vendors selling formally produced bottled water.

Informal water services can also operate completely disconnected from formal services, such as when the water source and all stages of the water supply chain are outside formal services.

3
Approaches to Improving Informal Water and Sanitation Services

Residents of informal settlements often do not benefit from large-scale investments aiming to improve a city's water and sanitation services. An investment project seeking to include the residents of informal settlements could develop an approach based on typical drivers, such as

(i)   the mandates, performance, and service targets of the government department, utility, or ministry seeking to improve the services;

(ii) the evidence-based sanitation and water needs, including the existing level of access to services, service providers' and residents' views, and gender and social inclusion needs;

(iii) the array and quality of services provided by intermediaries—informal businesses, nongovernment organizations (NGOs), and social enterprises—in the settlements; and

(iv) a mix of household and community services, dependent on the other drivers, residents' aspirations, and what is technically and financially feasible.

Private Household vs. Shared Community Service Delivery

The case studies and examples in this publication illustrate two modes of service delivery for water and sanitation in informal areas:

(i)   private household facilities, and

(ii) shared facilities at the community level.

In both modes (Box 1), separately or in combination and considering the appropriateness of either, an intermediary can overcome many barriers informal settlement residents face in accessing sustainable services (Box 3).

Private Household Services

Most families aspire to have private water and sanitation services for convenience, control of use and cost, and privacy. The household accepts responsibility for paying regular bills and operating and maintaining their facilities. The services are usually delivered as piped water within the household plot, either in the yard or the house, and a private latrine in the yard or the house.

Box 1: Summary of Household vs. Shared Services

Household-level services (private latrines and taps) are the aspirational level of service for most people in most places.

Community-level services (shared toilet facilities, tap stands, or water kiosks) may in some circumstances be acceptable, usually where household services are impossible. However, they require high levels of community engagement and substantial efforts to establish sustainable, professional operation and management, and efficient, convenient payment mechanisms.

Source: Asian Development Bank.

Shared Community Services

Where household-level services are not possible because of cost, lack of space in private households, or other constraints, shared community or public facilities, such as a communal toilet block, tap stands, or water kiosks, may be an acceptable improvement. However, they are unlikely to be at an aspirational service level. If households already have access to private facilities, even poor-quality services, it is unlikely they are willing to walk to use shared facilities regularly.

Where genuine community engagement and agreement exists, community services can be an accepted improvement over poor-quality informal services in some cultures and contexts. Community engagement is needed to understand land availability, power dynamics and conflicts, community needs and acceptance of services, and affordability.

Community toilets and water kiosks may also provide additional value and benefits, such as laundry facilities; the sale of soap and menstrual hygiene management products; and opportunities for employment, social interaction, and information dissemination. They can also be well designed and constructed to overcome challenging environments— for example, raised to avoid flooding or high groundwater.

Tap stand in an informal settlement. Residents collect and carry water from community taps stands in Port Moresby, Papua New Guinea (photo by Penny Dutton/WaterAid).

Similarly, community water standpipes can be planned to be within a short walking distance for most of the community and have good water pressure, availability, and quality.

However, simply building shared facilities is never an effective way to improve sanitation or water supply. Communal facilities will almost always require an intermediary or entity to pay water, electricity, and sewerage bills and to manage, operate, clean, and maintain the facilities. For communal taps and standpipes, a fair system of paying for water and maintenance is one of the most challenging aspects of their operation.

Failure to consider the community context and provide for sustainable operation and maintenance inevitably causes conflict, resulting in the facilities' disrepair and disuse.

Services for Informal Workers

Serving the needs of informal workers, businesses, and traders within formal or informal areas requires a similar evidence-based approach for design and planning, but often different solutions, partly because of the workers' mobility and scattered locations. Services must be responsive to their needs and meet citywide inclusive sanitation and human rights to water principles. Questions to consider when planning these services include the following:

(i)   What is the evidence of need and level of demand from the informal workers and traders in the areas? Are there nodes or hubs where workers congregate? Are there gender issues relevant for informal workers, such as female garment workers who need gender-sensitive toilets and menstrual hygiene management facilities?

(ii) What are the water and sanitation services available to informal businesses and workers? What are the gaps and failings in these services and their costs?

(iii) Are there pockets of communities lacking sanitation and water services in the same areas? If so, how would they be considered in the intervention?

(iv) Are there any nearby examples of well-functioning services, such as communal toilets operated under a performance contract (India) or public standpipes (proposed in Dili, Timor-Leste), that could be replicated or extended to these areas?

Intermediaries as Water and Sanitation Service Providers

For informal urban settlement residents in Bangladesh, Cambodia, India, Nepal, the Philippines, and Papua New Guinea, water and sanitation service delivery can involve small and medium-sized enterprises (SMEs) as intermediaries between them and the formal sector, providing either household or community facilities or both (Figure 5; Box 2). These intermediary service providers overcome barriers that poor, informal urban communities face in accessing formal services by innovating and taking risks (Box 3).

Figure 5: Role of Intermediaries in Providing Water and Sanitation Services

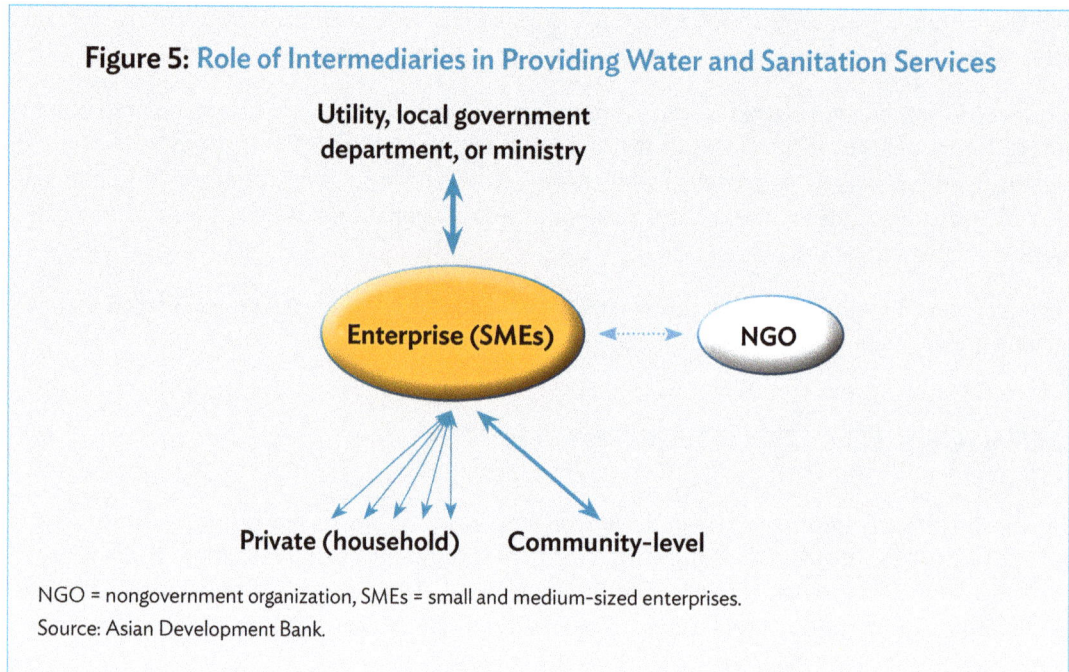

NGO = nongovernment organization, SMEs = small and medium-sized enterprises.
Source: Asian Development Bank.

Box 2: What Is an Intermediary?

A water and/or sanitation intermediary could be a for-profit small or medium business, a social enterprise (a for-profit part of a nongovernment organization) or formal community-based organizations or nongovernment organizations.

Intermediaries are organizations with the incentive to sustain their operations based on the ability to charge a fee-for service, employ management, technical and financial staff.

Source: Asian Development Bank.

Box 3: Barriers Overcome through Service Provision by an Intermediary

Private Services
- No service in the street
- No address for regular bills
- No land or house tenure
- Unaffordable connection fees

Community Services
- Tenant of rented room or house
- Landlord will not give permission
- Multistory dwelling with no service

Source: Asian Development Bank.

These SMEs range in legal status from semi-informal to formal businesses. They encompass small for-profit businesses, social enterprises, community-based organizations (CBOs), and NGOs. Unlike large utilities, SMEs can be flexible, creative, and innovative in response to the needs of particular informal settlements and can deliver decentralized private and community services supporting the achievement of the Sustainable Development Goals (SDGs). In this publication, "SMEs" is often used as shorthand for intermediaries.

Unequal Access to Financing

Unequal access to financial resources for capital infrastructure, operation and maintenance, and capacity development exists among water supply and sanitation service providers.

National funding sources (from domestic revenue or international loans repaid over time from domestic revenue) typically fund substantial capital investments. Governments usually source and pay for formal municipal (shared) infrastructure—typically through utilities—and households (users) pay for piping and plumbing on their private property. Households also pay for the operational costs of the services they receive through tariffs, fees, user charges, or subscriptions.

Intermediaries such as NGOs, CBOs, and small private businesses can rarely access international loans or large government funds. Instead, these smaller investments are funded by external grants, some of which are available through NGOs and foundations, but rarely on a national or city scale. Private sector SMEs have the least access to diverse funding sources and rely on service revenue to support operation and maintenance. Raising capital can be particularly challenging for small SMEs.

Table 5 shows the range of funding sources available to different types of water supply and sanitation service providers.

Table 5: Funding Sources for Water and Sanitation Service Providers

Financing Requirement	Government Departments and Utilities	Intermediaries	
		NGOs and CBOs	Private Sector (including social and small and medium-sized enterprises)
Capital infrastructure investments	International loans to central governments, national budgets from taxes, customer receipts, private capital	Local and international grants, often linked to an international or local NGO	Private loans, savings, profits, innovation grants
Management, operation, and maintenance of services	Customer tariffs and revenue, intergovernment transfers and subsidies	Customer fees, tariffs, and income from complementary services	Customer revenue, profits, diverse product offerings
Institutional and staff capacity building	National budgets, grants as part of international loans, international technical assistance projects	International or local NGO support, grants from foundations and institutes	Revenue and profits, technical assistance project implementation, research grants

CBO = community-based organization, NGO = nongovernment organization.
Source: Asian Development Bank.

Paving the way. Services provided by informal providers include water tankers, bottled water delivery, shared or community latrines, unregistered pit emptying, container-based sanitation, and piped water to a private household or shared tap (photo by Tubig Pag-Asa, Inc.).

4

Case Studies of Intermediaries in Sanitation and Water Supply

Range of Services Provided by Intermediaries

The case studies across six countries in Asia and the Pacific highlight how intermediaries address the service provision gap for informal settlements and formal areas (Table 1). Services include highly customer-focused additions beyond basic water and sanitation delivery, as described in this section and summarized in Table 6. Detailed information on each case study is provided in Appendix 1.

Sanitation Case Studies

The sanitation case studies provide examples of sanitation hubs, which are public or community facilities that include toilets, bathing, laundry, and water supply. Often, the hubs are women-centered and disability-friendly (Aerosan in Nepal, Suvidha in India, and Bhumijo in Bangladesh). When well-designed, they include good lighting, access ramps, female caretakers, dispensers for sanitary products, and diaper-changing tables.

Water availability is assured by sourcing from public piped supplies where possible (Bhumijo). Otherwise, water is delivered by tanker to storage (Bhumijo), or rainwater and recycled water are used (Aerosan).

Where feasible, the sanitation hubs and toilets are connected to sewers (Bhumijo); where this is not feasible, waste is treated on-site in septic tanks (Bhumijo) or biodigesters (Aerosan).

The Aerosan sanitation hubs in Nepal are environmentally friendly and include water saving, gray water recycling, rainwater harvesting, energy saving, and solar lighting. They are also kept clean to encourage use. This professional service extends to Aerosan's teams, referred to as "hygiene hosts"—a term with high status, similar to flight attendants.

Water Supply Case Studies

The water supply case studies illustrate services from SMEs that operate as "mini utilities" or delegated management entities selling utility water and providing water connections, network operation and maintenance, billing and revenue collection (Shobar Jonno Pani [SJP], Tubig Pag-Asa, Inc. [TPA], Motu Koita Assembly [MKA]).[9] The 1001fontaines case study shows the development of water sources, extraction, water treatment, and operation and revenue collection independent of utilities. The service providers in the case studies produce and deliver

---

[9]   The MKA case study in Papua New Guinea is under development; MKA planned to begin supplying water from kiosks in 2023.

Table 6: Range of Services Provided in Case Study Examples

	Sanitation and Water Supply Services Provided	
Case Study	Informal Settlements	Mobile Informal Workers
Bhumijo Social enterprise	2 public toilets, including: - night lighting - extended operating hours - water collection points - MHM products - diaper-changing facilities - user-feedback device - footfall-counting devices	26 public toilets for informal workers and traders, customers in bazaars, kitchen markets, and wet markets
Suvidha Partnership	7 women and disability-centric sanitation centers, including: - toilets and showers - night lighting - sanitary products - diaper-changing facilities - laundromat with washing machines - water supply	n/a
Aerosan Social enterprise	5 public toilet blocks, including: - women and disability-centric design - wi-fi - after-dark sensor lighting - non-touch taps and urinals - air hand drying - sanitary product dispenser and pad disposal - diaper-changing facilities	3 public toilets serving taxi drivers, market vendors, and nearby informal settlement areas
Laguna Water A joint venture of PGL and Manila Water	• Water supply • Sewage management • Desludging of septic tanks • Communal toilets • Portable toilet solutions	n/a
TPA Social business	• Household piped water • Communal standpipes • Private household toilets • Firefighting capacity	n/a
SJP Social enterprise	• Household piped water • Communal standpipes • Household latrines	n/a
1001fontaines Social enterprise working with local entrepreneurs	20-liter bottled drinking water with: - home delivery service - quality guarantee	n/a
Motu Koita Assembly Social business	Water kiosks with: - disability access - information and payment hubs - home deliveries - stores for sanitary items Household piped water (to be developed)	Water kiosk at settlement area outside village

MHM = menstrual hygiene management, n/a = not applicable, PGL = Provincial Government of Laguna, SJP = Shobar Jonno Pani, TPA = Tubig Pag-Asa, Inc.
Source: Asian Development Bank.

consistently high-quality drinking water, by using treated utility water, adding chlorine treatment to guarantee quality water (to a better standard than the utility supplying the water) (TPA), or treating and testing water monthly in the NGO's laboratory and biannually by the Cambodian Ministry of Mines and Energy (1001fontaines).

Operational Features of Intermediary Services

The case studies share several common features in how intermediaries operate and approach service delivery.

Formal Government Relationships

The intermediary organizations have strong formal ties with local government and utility services. These relationships are based on memorandums of agreement or understanding and include formal business registrations, operating permits, and a clear allocation of responsibilities. Cordial relationships with local governments and utilities are essential for approving land for communal, shared, or public facilities and preventing obstruction of operations. These relationships fundamentally differ from informal service providers operating without the formal knowledge or approval of local authorities.

Parent Organization Relationships

Where the intermediaries have an established connection to a parent organization, such as an international or national NGO (Bangladesh, Cambodia, Papua New Guinea, and the Philippines), the parent organization can provide support and add value to the services offered. Parent organizations can offer orientation on the service delivery model, critical set-up support and capital costs, ongoing capacity building, networking among similar organizations, and specialist technical advice. The connection of parent organizations to international networks and donors can also be an advantage in raising funding for capital works and emergency funding, such as for subsidies for free water allocations during the COVID-19 pandemic in Bangladesh and the Philippines.

This contrasts with Inpart Waterworks and Development Company (IWADCO)—an independent local water vendor, not supported by any international or local NGO, on-selling utility water to urban poor populations in Manila, Philippines from 1997 to 2013. IWADCO struggled to raise capital and access credit for a piped network because of perceptions that it was not an official utility and, thus, a high-risk investment and that providing water to people living below the poverty line was unprofitable.

Community Engagement

Intermediaries can potentially create meaningful engagement and long-term relationships with communities. The ongoing physical presence of social enterprises in a community and regular meetings and community interaction build social capital and trust. These relationships are the basis for the "add-on" social services communities need, such as laundry facilities, firefighting capabilities, and school drinking water. Intermediaries also employ members of the community.

During the COVID-19 pandemic, intermediaries provided critical communication links between local authorities and the community. In the case of TPA in the Philippines, its essential service providers were the only connection informal settlement residents had to the outside world during mandatory lockdowns. Water and sanitation intermediaries were also critical in distributing hygiene infrastructure and materials—including handwashing stations, water, soap, sanitary pads, and masks—and awareness messages on the status of services and the importance of handwashing and social distancing.

Capacity Building

Ongoing capacity building is a feature of the case studies, particularly where parent organizations—such as international NGOs Eau et Vie, 1001fontaines, Toilet Board Coalition (TBC), and WaterAid—have a strong continued presence or links to research organizations. For example, when establishing itself in a new country, 1001fontaines provides intense capacity building during the first year to build a national organization that delivers support to entrepreneurs through regional and national teams. It also provides ongoing support to enable the national organization to manage the network and specialist technical support, for example, in climate change. Eau et Vie brings valuable technical expertise in water, sanitation, hygiene, and water quality treatment to Bangladesh. WaterAid has assisted MKA with water supply governance arrangements, including terms of reference, agreements, and financial models. The sanitation case studies also feature additional ongoing learning, sharing, and capacity development support.

Financial Models and Viability

All service providers in the case studies charge user fees for their services, with some flexibility for low-income households and individuals. The case studies show that customers are willing to pay for better, safer, cleaner, and more reliable services—at affordable prices—in place of free, poor quality, unreliable, inconvenient, or more expensive alternatives (WaterAid 2016).

Government water and sanitation systems in most Asian and many OECD countries fully or mostly subsidize capital investment costs and may contribute to operational costs as a public good. By comparison, the private sector and NGOs have fewer and less-secure options for operation and maintenance costs that enable efficiency, good billing and collection, and high-quality services for which customers are willing to pay.

Yet, for services to be well-maintained and viable long-term, the combined income should cover all operational costs and, especially for the private sector, return a reasonable profit. The case study examples conformed to this model, with external funds providing the capital infrastructure and training costs and users paying a service fee to cover management and other operational costs. Additional external capital funding may allow for further investment in climate change and resilience interventions (guttering and rainwater tanks, gray water recycling, and sensor taps to reduce water consumption), which reduce operational costs and investment in expanding or upgrading systems.

Aerosan in Nepal operates a café near its toilet facility that uses the biogas from the toilet block for energy, and the café's profits contribute to the toilet block's operational budget. Aerosan estimates that it takes from 3 to 6 months for the facility to become self-sustaining, and during the first 3 months, Aerosan supplements the operational budget. Bhumijo sells advertising space to supplement user fees to cover operation and maintenance costs. MKA has underwritten establishment costs for the new Pari Water Services, such as staff salaries, bulk water charges, and contributions toward a new pipeline, while revenue is generated and accumulated from the services to pay for these costs.

Advantages of Intermediary Services

The case studies reveal several common advantages of intermediaries' sanitation and water service delivery.

Quality Services

All SMEs in the case studies deliver high-quality and reliable services such as continuously clean, safe, and functioning toilets and 24/7 safe water supply. This attention to quality ensures facilities and services are used and paid for. All the enterprises emphasized the importance of their brand as being trusted, reliable, and offering high-quality services. The continuation of services throughout the COVID-19 pandemic only furthered this trust.

Meeting Needs of the Poorest Households

The SMEs have developed local payment structures and fees to meet the incomes and affordability of the community and the poorest households. Although tariffs for water supply are more expensive than utility water, the adaptability and flexibility of payment systems—such as allowing small regular (daily) water payments—enable poor households to manage cash flows. Subsidies for people living in extreme poverty (TPA) or pensioners (Laguna Water) also show responsiveness to the needs of very low-income households. During the COVID-19 pandemic, TPA implemented a policy for flexible arrears payments to avoid crippling poor households with debt and maintain water connections.

Innovation for Efficiency and Resilience

The SMEs exemplify a readiness to adapt and innovate their services when it leads to efficiencies and streamlining. Larger SMEs, including Bhumijo and Laguna Water, have innovative research and development teams to improve their products and services. The use of technology, such as digital payment systems, is a feature of water and sanitation services in the case studies. The Papua New Guinea model goes beyond mobile payments to include asset management and a ticket system for reporting issues, e.g., billing disputes, requests for a household service line, leaking pipe, and notifications for action and issuing purchase orders. In addition, Pari village in Port Moresby has adapted TPA's approach to pipeline laying in informal settlements and will use large conduits to accommodate future multiple service lines. This approach offsets connection costs because ground disturbance and excavation are minimized when future connections are made.

The case studies also demonstrate SMEs' ability to build resilience through their approaches. In Papua New Guinea, the utility's water supply network cannot reliably provide 24/7 water. The urban village water supply system under development includes 9,000-liter overhead storage tanks at each kiosk to create resilience during utility downtimes. The tanks also support the sustainability of the "user pays" model—if there is no water, there is nothing to sell. The Papua New Guinea model includes several metering levels—bulk meter, storage tank, retail point—to measure sales and water losses, in part to overcome the social challenge of kiosk operators' pressure from family members to give them water for free. In India, Suvidha has adopted rainwater harvesting to minimize dependence on external water sources.

Responding to Emergencies

The ability of SMEs to respond to unforeseen external events is evident from their adaptability during the COVID-19 pandemic. SMEs rapidly adopted innovations that supported the continuation of services during challenging periods of lockdown. The sanitation service providers introduced online and digital payment methods, including radio-frequency identification cards, QR codes, and mobile app-based payments. Sanitation SMEs installed user-feedback and footfall-counting devices to facilitate remote monitoring and inform cleaning schedules (Bhumijo, Aerosan).

The pandemic also accelerated investment in smart and new hygiene technology, such as air hand dryers, sensors, pedal-operated taps for handwashing, pedal-operated flushing, sensor-based urinals, and diaper-changing stations. Bhumijo developed and sold these devices to local governments, businesses, and development agencies, with sales helping to stabilize Bhumijo's reduced cash flow because of the pandemic.

SMEs also innovated by using technology to maintain communication during the COVID-19 pandemic. All the SMEs reported using multiple platforms to communicate between management and staff, including WhatsApp, MS Teams, Viber, Telegram, Messenger, Zoom, and mobile calls and texts. Customer communication expanded to Facebook, messaging via Barangay Information channels (Laguna), Messenger, and office mobile phones (Bhumijo, TPA).

Water supply delivery adaptations were driven by the need to continue essential services under social distancing and lockdowns. For example, meter readings ceased, a phone hotline was installed to report emergency repairs, and payment arrears policies were adapted to extend repayment time (SJP and TPA).

Desludging septic tanks was not considered an essential service in the Philippines during the COVID-19 pandemic. Such services were paused during the strictest COVID-19 pandemic lockdown in 2020 until the government secured an exception.

Economic and Health Benefits

Interventions by the SMEs have demonstrated economic and health benefits to informal settlement residents. The case studies indicate that the enterprises boosted the local economy through job creation, reduced costs for water provision and facility access, and improved time savings, particularly for women. In Nepal, Aerosan offers free access to public toilet hubs for people with disabilities, older people, and children—as well as anyone else unable to pay. The SMEs also promoted public health through providing potable water, increasing water availability, reducing the need for open defecation with the introduction of public toilet facilities, and ensuring uninterrupted water and sanitation services during the COVID-19 pandemic. In Bangladesh and the Philippines, SMEs offered water relief programs during the pandemic to reduce the spread of the disease.

Climate Change Adaptation

Several SMEs are building climate change adaptation approaches into their services. Even on a small scale, investing in gray water recycling systems and solar panels lowers operational costs and increases sustainability if these interventions are well maintained. To ensure value for money, investment costs must be balanced against technical requirements, low running costs, robustness (durability), and ease of operations. Some examples of adaptation and water and energy efficiency approaches reported by the SMEs include the following:[10]

Aerosan

(i)   Rainwater harvesting;

(ii)   Sensor-based automated water system to conserve water use;

(iii)   Biodigester to treat fecal waste, helping prevent human waste from entering water bodies;

(iv)   Methane gas from biodigester used as fuel source in a café;

(v)   Gray water recycled and used for toilet flushing;

(vi)   Solar lighting to minimize electricity use; and

(vii)   Chemicals avoided in toilet cleaning to support biodigester and sludge reuse.

Suvidha Sanitations Hubs

(i)   Suvidha centers have saved 35 million liters of water since 2016, based on a 2021 impact assessment study by RTI International (TBC 2021).

(ii)   Each center saves 4 million liters of water annually using a closed-loop approach to reusing water from handwashing and laundry facilities. Water from handwashing and washing machines is treated and reused for toilets.

(iii)   The center in Ghatkopar, India treats and reuses gray and black water at a community-toilet level. This is achieved by a hybrid treatment system based on a decentralized wastewater treatment system (DEWATS) approach.[11]

1001fontaines

1001fontaines estimates 17,000 tons of carbon dioxide emissions have been averted annually by using solar power for water treatment at the kiosks, compared to household water treatment by boiling.

Capacity for Scale

Some intermediary SMEs can deliver quality, clean, reliable, sustainable, and affordable water and sanitation services to an entire community, as well as additional communities.

While the case studies considered vary by number of customers served daily, all the intermediaries have actual or potential ability to scale. For sanitation, Aerosan in Nepal has eight locations averaging 800 customers daily; Suvidha in India operates toilet complexes in seven locations reaching 550 people daily and plans to add more; and Bhumijo in Bangladesh has 15 locations serving 6,000 people daily. Laguna Water in the Philippines has 116,000 water customers across three cities, which gives potential to scale sewerage and fecal sludge management (FSM) services to more customers.

---

[10] Climate impacts and water and energy efficiencies are as reported by SMEs. The claims have not been independently verified regarding quantity and sustainability.

[11] DEWATS refers to a decentralized, community-level wastewater treatment technology. The passive design uses physical and biological treatment mechanisms to treat both domestic and industrial wastewater sources. DEWATS is designed to be affordable and low maintenance, use local materials, and meet environmental laws and regulations.

For water, TPA under Eau et Vie in the Philippines operates in 18 communities serving 6,058 households; SJP in Bangladesh has two locations (Chattogram and Bhashantek) serving 2,200 low-income households. 1001fontaines in Cambodia has 270 water kiosk services throughout the country and has developed urban models for Madagascar and Viet Nam.

The key to scaling up is a sustainable and financially viable model. For example, 1001fontaines knows exactly how much water must be produced and sold daily to cover each kiosk's basic operating costs. Similarly, Eau et Vie has a proven financial model for covering operating costs that can be tailored to other locations.

The Utility as Service Provider

An intermediary is not always needed, and in some cities, water and sewerage services can be provided to low-income areas and informal settlements directly by utilities, e.g., Laguna Water, Fiji Water Authority, Manila Water, Maynilad Water Services, and Kolkata Municipal Corporation. Examples from Africa demonstrate service implementation through dedicated low-income settlement units (e.g., Lusaka Water and Sewerage Company, Zambia; and National Water and Sewerage Corporation, Uganda) or advisory units for low-income communities or settlements (e.g., Nairobi City Water and Sewerage Company, Kenya; Dar es Salaam Water Supply and Sanitation Authority, Tanzania; and Ghana Water Company Limited, Ghana) (Peal and Drabble 2015).

Utilities can provide water and sanitation services in informal settlements directly to households or to community or shared facilities (Figure 6).

Figure 6: Water and Sanitation Services Provided by a Utility

Utility, local government department, or ministry

Private (household)   Community-level

Source: Asian Development Bank.

For example, the Water Authority of Fiji (WAF) is authorized, but not obliged to serve all communities in Fiji—including informal and peri-urban areas. WAF views informal settlement residents as potential customers, tracks their numbers, and has developed ways to work around land tenure complications to serve them. If customers do not have land tenure documents, WAF allows customers to instead submit government-issued identification credentials. WAF approves connection requests to illegal settlers on private land with authorization from landholders or owners (Schrecongost and Wong 2015).

WAF also has an innovative approach that allows it to serve informal settlements while reducing its infrastructure investment risk. WAF does not invest in distribution networking in informal settlements. Instead, approved customer meters are installed at the edge of the settlement, and the households install flexible distribution piping (typically polyvinyl chloride pipes) from the meter to their home. If the distribution piping crosses others' property, customers must provide letters granting a right of way from those households. This solution allows the utility to reach informal settlement households with relatively low risk.

Water as revenue. Informal settlement residents represent important additional revenue for water utilities and are connected by innovative service delivery and expansion (photo by 1001fontaines).

However, this model has some weaknesses. Meters on the main road are not protected and are sometimes vandalized or damaged. WAF does not assist with purchasing, placing, or protecting distribution pipes. Pipes often run through pathways and contaminated drains, resulting in damage and subsequent contamination. When water pressure is high, damaged pipes leak, causing standing water and water losses that inflate water bills. High-density polyethylene pipes are slightly more expensive, but better suited for settlement conditions.

A more comprehensive utility approach is Manila Water's Tubig Para Sa Barangay program. Under the program, Manila Water connects low-income households to the utility's piped water supply by introducing installment payments for connection charges and subsidizing connection costs. Previously the high one-off connection fee had been a barrier to connecting. The program has made intermediaries such as Inpart Waterworks and Development Company (IWADCO)—a private water vending and on-seller company—redundant. The utility supplies household water at an affordable price and undertakes functions such as pipe laying, connections, maintenance, and meter reading in the informal communities. The program grew out of Manila Water's target to achieve connection goals, a corporate social responsibility program, and a desire to reduce illegal connections contributing to nonrevenue water. To overcome the lack of access to land tenure, Manila Water developed agreements with local governments, allowing connection forms to state that the connection to the utility would not be taken as a sign of landownership.

Additionally, the *barangay* (local ward) office issues a certificate vouching that the applicant is a resident of the area. This discourages people from moving to the area to exploit the water supply. A subsequent agreement between the government and Manila Water has formalized the subsidized connection fee and operational cash flow. The other Manila concessionaire, Maynilad Water Services, operates a similar program.

There are three critical requirements for effective service delivery at scale by utilities to informal settlements and low-income communities:

(i) **Corporate commitment.** Top management must support service delivery for informal settlements and low-income communities and integrate this into the rest of the organization.

(ii) **Well-defined roles and responsibilities.** Roles and responsibilities for those delivering the strategy must be clear, consistent with other departments, and understood throughout the utility and by stakeholders.

(iii) **Clear plans and performance indicators.** Short-, medium-, and long-term objectives for serving low-income communities must be clear and integrated with the utility's wider strategies. Key performance indicators must include metrics to assess (a) contributions of low-income areas to total revenues, (b) nonrevenue water, and (c) consumer satisfaction.

Other important features include adequate resources and power of the responsible division to drive the approach and a balance of technical and community staff (Peal and Drabble 2015; WaterAid 2009).

5
Planning Investment Projects That Include Informal Settlements

This section provides an outline planning guide and suggestions for developing a project component for providing water and sanitation services in informal settlements, focusing on delivery by intermediaries. The context for these suggestions is for such an initiative to be a part of a substantial city-based water and sanitation investment project rather than a small stand-alone initiative.

Benefits of Working with Intermediaries

The case studies and examples demonstrate that services in informal settlements provided by intermediaries can yield significant benefits, including the following:

Reliable service provision. Dependable, clean, and safe services are supplied to residents excluded from formal services due to lack of land and house tenure, lack of building space, or low income.

Public health benefits. Public health improves by reducing open defecation; addressing the lack of potable water; increasing water availability for bathing, laundry, and handwashing; and providing hygienic water storage.

Improved safety and time savings for women. Dignity and security for women and girls are promoted by providing safe and affordable alternatives to open defecation and unhygienic facilities. In-house or nearby water supplies promote hygienic household water management and relieve women in particular of the time-consuming and costly chore of collecting water from distant, unreliable, or crowded sources. Clean, private, and convenient toilets also enable dignified, hygienic menstrual hygiene management. Access to affordable washing machines in community sanitation centers further reduces work for women, who typically bear the burden of most household work (Suvidha).

Positive social outcomes. Organized, equitable, and fairly priced water services promote positive social outcomes, such as reduced community conflict over water (TPA), suppression of criminal networks exploiting the water market (SJP), better value for money (Motu Koita Assembly [MKA]), and reduced vandalism of water infrastructure (TPA, SJP, MKA).

Environmental benefits. Services result in a cleaner environment and help address or reduce climate impacts through water and energy efficiencies.

Community engagement and trust. Community engagement and trust are improved through decentralized, resilient, locally managed services that are customer-focused and provide clear communications.

Employment creation. Local jobs are created for marginalized community members, boosting the local economy. In Cambodia, 1001fontaines employs more than 900 people (about three per service); in Bangladesh, SJP employs 98% of staff from the local areas (50% from within the intervention areas), 35% of whom are women. Intermediaries in Bangladesh and the Philippines create local opportunities for plumbers, bill collectors, community organizers, caretakers, engineers, branch managers, and field coordinators.

Sanitation-plus and add-on services. Many more add-on services are possible than typical utility-style water and sanitation services, which may help support financial viability. With support from their partners, intermediaries can offer a platform for complementary actions such as hygiene promotion (all case studies), sanitation improvements (SJP), firefighting and emergency management (TPA), solid waste management, laundry facility upgrades (Sudhiva), and the sale of menstrual hygiene management and other sanitary products (Sudhiva, Aerosan, Bhumijo).

Where Could an Intermediary Model Be Suitable?

The first option for serving informal settlements is for an existing utility to deliver the services. The policies and approaches of utilities such as Phnom Penh Water and Sewerage Authority in Cambodia, Manila Water and Maynilad Water Services in the Philippines, WAF in Fiji, and Kolkata Municipal Corporation in India demonstrate that this is feasible and sustainable (section 4.4).

However, where a utility has adequate water resources and treatment capacity, but is unable or unwilling to serve informal settlements directly, working with an intermediary could be considered if the following conditions are met:

(i)   There are unserved or poorly served informal settlements within or near areas where investments are planned to expand water and sanitation infrastructure and services.

(ii) A utility (or local government) is willing and motivated to work with nongovernment organizations (NGOs), community-based organizations (CBOs), or social enterprises or can be readily convinced to try.

(iii) Suitable intermediary organizations are active in the same city, country, or region.

(iv) The legal and regulatory framework is not a barrier, and suitable legal mechanisms such as memorandums of agreement or understanding and contracts are available.

It is also helpful if the utility (or local government) has appropriate performance drivers, e.g., reducing nonrevenue water, increasing or expanding the customer base, or achieving national or local water and sanitation access targets such as Sustainable Development Goals (SDGs). If these are not already in place, they can be developed.

Are Household or Community Services More Suitable?

As explained in section 4, improved services may be provided at the household or community level. In some communities, these services can be provided at both levels, encompassing the poorest tenants of rental rooms and less-poor residents who may be owner-occupiers. Household services are usually preferred, but are not always technically feasible, sustainable, or affordable.

Every informal settlement is different, and the potential for various types of services must be assessed based on the settlement's features and city context before the most suitable model is selected. Data on the following factors should be obtained and used as evidence to support the choice of model:

(i) **Community basics.** Information should include the population, number of households, household size, population density, land and house tenure, community composition and cohesion, leadership structure, employment rates, household income, geography, and topography.

(ii) **Water and sanitation arrangements.** The type and cost of water and sanitation services and users' satisfaction level with these services should be assessed. Are these shared or household services? The improved services must be of better quality and better value for money.

(iii) **Access to infrastructure and sources.** Research should encompass access to water and sewerage infrastructure. Are there water or sewerage pipes nearby? Are there alternative water sources that could be developed? What is the distance to sewage or fecal sludge treatment plants?

(iv) **Community aspirations.** Substantial demand for improved water and sanitation services among informal settlement residents should be demonstrated. Which service improvements do residents want (e.g., quality, availability, convenience, or cost reduction)? What improvements are residents willing to pay for? How many people are willing to pay more? Improved services should not be launched in places where households are satisfied with existing services or where the utility or municipality is better placed and willing to provide formal services. Planning for shared facilities in low-density areas where people do not want to walk long distances should also be avoided (Box 4).

Box 4: Household or Community Toilets?

ADB Karnataka Integrated Urban Water Management Investment Program

The initial concept of the Asian Development Bank Karnataka Integrated Urban Water Management Investment Program was to provide individual toilets for some households and community toilets for people without access to toilets using output-based financing. However, during implementation, demand was identified for individual toilets and toilets in girls' schools and colleges, leading to a change in approach. Local nongovernment organizations, which were better connected to the communities, conducted a needs assessment and identified the necessity for a revised plan. They also conducted behavior change training and sanitation marketing to stimulate demand and verified the outputs. The number of toilets required and their locations were identified and approved by the community. The design and supervision consultants supported construction alongside community-based groups (under the guidance of nongovernment organizations). Weekly progress review meetings were conducted on grant-related work. The project implementation unit and project management unit were available for coordination with the education department and school management personnel, and payment to the contractor was based on the number of toilets constructed.

Source: Asian Development Bank.

(v) **Stakeholder analysis.** Research should include information on the formal and informal service providers and what they provide. What are the barriers and opportunities for their services? Who are the potential intermediaries?

Scope of Work

When the government department or utility decides to take an intermediary approach, a scope of work (or terms of reference) is required. Based on lessons from the case studies, the scope of work for an intermediary could include the following elements:

(i) **Baseline data.** Establish a baseline for the community through a household survey. Depending on the informal settlement's size, this could be a census of all households or a sample survey to understand their situation, including preferences, needs, affordability, and vulnerability. The survey may be the only data source for the community, especially if informal settlements have not been included in government censuses or other surveys or databases. The survey process can also serve as an opportunity for sharing information with the community.

(ii) **Inclusive approach.** Community engagement and participation aspects must prioritize gender equality and social inclusion in both the methods—such as ensuring women and marginalized groups (e.g., people with disabilities and ethnic minorities) are included in planning—and in the outcomes, with services reaching the most marginalized populations.

(iii) **Continuous coordination.** Ongoing, close consultation should be maintained with the following stakeholders:

    (a) community members to identify and develop appropriate service levels and ensure model is designed based on what people want and are willing to pay for, balanced with technical feasibility and affordability;

    (b) community leaders to support them to inform, advocate, and mobilize residents;

    (c) local partners such as civil society organizations and NGOs to help support community engagement and service provision; and

    (d) utilities, city authorities, and ministries (as relevant) and their designated project consultants and contractors to support stakeholder buy-in for well-coordinated implementation and management.

(iv) **Integration.** Integrate the informal services in a wide city investment project with the utility, local government, or ministry client. The timing of the informal settlement component development should optimize access to new municipal water or sewerage systems whenever possible. Aligning informal settlements with formal services is the most cost-effective approach. Retrofitting informal settlements later is a costly missed opportunity.

(v) **Clear responsibilities.** Clarify institutional roles and responsibilities, including

    (a) legal mandates (in which context are agreements needed, for which services, and between which parties);

    (b) tariff setting for wholesale and retail;

    (c) operation and maintenance, including routine operations, maintenance, and repairs and long-term asset replacement;

    (d) billing, payments, and customer engagement; and

    (e) supervision, oversight, and reporting requirements.

(vi) **Asset assessment.** Assess asset ownership (e.g., for pipes and shared facilities) and handover procedure for assets; review landownership implications—who owns the land on which assets sit, easements, and instruments for settlement landowner contributions.

(vii) **Financial plan.** Develop a credible financial plan to achieve operational sustainability 6 to 12 months after commissioning (or during the second year of the project) based on recovering all operational costs from locally affordable and equitable user fees and encouraging the development of additional revenue sources from "add-on" services. An operational support subsidy is necessary for the first 1–6 months to enable the intermediary to cover operational costs as customer bases are built.

(viii) **Awareness-raising.** Use awareness-raising and behavior change communications approaches to address issues such as handwashing with soap, water consumption, tariff payments, and sanitation. Outline the approach and target audience and behaviors.

(ix) **Construction design.** Design and contract construction of the necessary infrastructure for the sanitation or water services based on utility and community discussions, topography, and local technical options. This phase includes preparing contract documents and tender processes following utility and development finance institution (DFI) rules.

(x) **Digitalization.** Integrate up-to-date local information technology (IT) skills to set up and operate a cost-effective and robust digital community service payment system with flexible payment options. Generally, cash payments should be minimized to enhance accountability and security.

(xi) **Medium-term support.** Provide medium-term backstopping and capacity-building support to develop local staff skills and ensure ongoing services, management and supervision, and repair skills. Assist with troubleshooting and technical support, periodic monitoring and analysis, refining business processes, and updating the financial plan.

Contracting Intermediaries

Where there are no services to informal settlements or the intermediary and utility are new to the approach, long lead times are required to establish relationships with the local government or the utility and credibility with the communities, as well as to conduct surveys and develop an appropriate model with agreed-on roles and responsibilities. For example, due to the new approach used for the Pari village water supply in Papua New Guinea, it has taken more than 2 years since the initial concept note for the intermediary and the utility to negotiate roles and responsibilities and formalize these in a memorandum of agreement. Rolling out the model to other urban villages is expected to be much quicker, as relationships and processes have already been established. Service agreements with the utility can be modified for a new setting rather than developed from scratch.

Where there are preexisting local connections and relationships with the utility and local government and experience with appropriate models, the case studies suggest a 4-year contract in two phases can be considered. The work will initially be intensive and may be full-time (up to 1 year), followed by 3 years of limited support and monitoring (Figure 7).

(i) **Phase one.** Moving from initial stakeholder engagement to training and commissioning, the intensive first year's work may include 3–4 months for studies, community engagement, and design; about 3–6 months for construction (Aerosan, Bhumijo, Suvidha) and technical setup; and 3 months for operational staff training and final commissioning. This setup

Figure 7: Estimated Project Time Frames

Years	Design and Set-Up		Monitoring and Support			
	1	2	3	4	5
Community engagement, design (3–4 months)	▉					
Construction (3–6 months)	▉					
Training, commissioning (2–3 months)	▉					
Monitoring (financial, technical, sustainability)		▌ ▌	▌ ▌	▌ ▌		▌
Technical support, troubleshooting		▉▉▉▉▉▉▉▉▉▉▉▉▉				
Organizational support, capacity development		▉▉▉▉▉▉▉▉▉▉▉				

Source: Asian Development Bank.

period can be reduced when the organization has adequate experience. Most case studies estimated 6 months could be enough; however, this will depend on the context.

(ii)   **Phase two.** The monitoring and support period of 3–4 years will include service delivery monitoring, financial model monitoring, ongoing liaising with the utility or local government, regular staff training and human resources support, supervision visits, and support during emergencies and for unexpected major repairs. Ideally, this phase allows for cost-effective support to several facilities within the same city or area with a year-on-year reduction. Monitoring and support are recommended for at least 2 years after the system has become financially viable.

The payment schedule for an intermediary should reflect the sustainability focus and supply the service provider with the working capital needed and incentive to deliver the outputs. It must also recognize which activities incur significant costs. These include the community survey, design and construction of the infrastructure, and monitoring and support, which should be phased out over time. For guidance, a payment schedule for a 4-year implementation period could comprise the following:

(i)   **Year 1: 60%.** This payment could be divided into tranches linked to community and stakeholder engagement (10%); technical design and staffing and financial plan (10%); infrastructure construction and IT systems (30%); and commissioning and documentation, training and capacity development, and behavior change communications (10%).

(ii)   **Year 2: 20%.** This payment is based on achieving financial sustainability to cover operating expenditures from user fees and other sustainable income sources. Evidence of performance may come from quarterly reports covering capacity building and support provided, user numbers, financial reports (profit and loss), adjustments, and lessons learned.

(iii)   **Years 3 and 4: 10%.** This payment is based on two 6-monthly monitoring reports of user numbers and financial records and a final handover report outlining the model's sustainability.

What Makes a Good Intermediary Organization?

Given the scope of work expected and lessons from case studies, the following characteristics are a guide for what to look for in an intermediary:

(i)   Locally based and working in informal settlements, preferably in infrastructure service delivery.

(ii)   Flexible and adaptable, entrepreneurial, creative, and open-minded. Able to adapt the model as information becomes available and able to communicate the concept to all stakeholders throughout the process.

(iii)   A track record of effective community engagement. Known and trusted by the community or similar communities.

(iv)   Good working relationships with local government and water or sanitation utilities.

(v)   Technically competent in designing, contracting, and maintaining water and sanitation infrastructure, with the skills to subcontract and supervise or a trusted partner to do so.

(vi)   Driven primarily by social objectives rather than publicity or academic or financial motives.

(vii)   Focused on long-term financial viability and realistic about the need for users to pay the operation, maintenance, and management costs—as opposed to a "charity" mindset.

(viii) Financially and administratively well-equipped and competent. Confident in using digital planning, survey, financial tools, and payment methods and able to engage others for specialized technical tasks.

(ix)   A capacity-building approach to designing and delivering appropriate training and long-term skills transfer.

(x)   Protects staff welfare and rights through adherence to legal protection and core labor standards.

(xi)   Good managers for a multidisciplinary service delivery model and ability to keep project momentum on track throughout planning, construction, operation, and monitoring phases.

(xii)   A formal "parent" relationship with a national, regional, or international organization that could help support capacity building and provide access to networking, capital, and other added value, such as monitoring.

The scope of work required is unsuitable for a typical water supply or wastewater design consultant and construction contractor, despite the subsidiary hardware components.

Managing Challenges and Risks

The case studies highlight real-world challenges and risks experienced by service providers in informal settings. These include those within the organizations, beyond their direct influence, and related to scaling up their operations.

Internal to Intermediaries

The financial viability of social enterprises is threatened by weak or uncertain financial sustainability and recovery of operational costs. For water supply, this starts with having a fair or discounted bulk water rate, to which the social enterprise must add their staffing and overheads.

Many informal service providers rely on support for financing capital investments, capacity building, and ongoing technical support. Support from international NGOs and philanthropic donors is often tied to project funding cycles or dependent on fluctuating trends and, hence, is less secure and in smaller amounts than the loan support from development banks that larger state-run utilities can access. In the case of water supply, ongoing support may be provided by subsidized international NGOs (Eau et Vie, 1001fontaines).

Providing sufficient committed staff with various skills is a typical internal challenge. Ongoing capacity building is needed to develop the required skills for service sustainability. It may be necessary to partner with other organizations to build these capacities over time.

External Factors

The COVID-19 pandemic provided a substantial external threat and rich lessons about the fragility and strengths of working with small and medium-sized enterprises (SMEs) as intermediary service providers.

Climate change and natural hazards threaten water and sanitation infrastructure and housing. Examples include cyclones, severe flooding, drought, earthquakes in the Philippines, and landslides and earthquakes in Bangladesh. Climate-related water resource depletion threatens the quantity of water utility supply, and shortages have knock-on effects on social enterprises' services.

Informal settlements remain vulnerable to changing political stances, with the continual threat of evictions in many places or selective recognition of settlements. Political conflicts and souring relationships with local government are other potential challenges, as well as the toll on relationships caused by elections and changing leaders. These can only be dealt with on a case-by-case basis. However, they are often navigable with sensitivity, a good understanding of the local political economy, and community support.

Stringent government regulations may also adversely impact financial sustainability. However, there are work-arounds to government regulations, such as providing services to the settlement boundary or stating that service access does not legitimize landownership (section 4.4).

Within communities, vandalism, infrastructure theft, intimidation, and standover tactics (mentioned in the Tubig Pag-Asa, Inc., Shobar Jonno Pani, and Motu Koita Assembly case studies) harm emerging social enterprises. However, most social enterprises reported that local connections and relationships with community leaders and members, although challenging, helped manage these threats. Incorporating monopolistic water vendors or informal pit emptiers into the new model, through employment, is another approach to tempering the threats.

Development finance institution (DFI) challenges working with intermediaries include

(i)   limited understanding and commitment to the model by project officers,

(ii)   procurement processes that are overly demanding of small intermediaries,

(iii) DFI project time frames that may not provide enough post-construction support, and

(iv)   safeguards requirements such as land acquisition.

Examples of risks and challenges for sanitation intermediaries and possible mitigation measures are provided in Table 7.

Table 7: Sanitation Business Challenges and Mitigation Measures

Intermediary Challenges and Risks	Mitigation
Single-focus intermediaries who cannot survive with reduced customer numbers, changes in local population characteristics, or severe climate change impacts.	Encourage businesses to operate in multiple settings (or plan to do so), and/or offer complementary services and products that are more resilient to system shocks, such as pandemics. Offer a mix of community and public toilets rather than specializing solely in community toilets or public toilets.
Unregulated services result in low standards, poor quality construction, and inappropriate allocation or payment contracts to operate facilities.	Advocate for standards and regulations on the construction, operation, and management of sanitation facilities and services, with realistic incentives for complying. Propose performance-based management contracts that align with standards and regulations while providing room for innovation, business development, and growth.
Closure of sanitation facilities during national emergencies.	Advocate for all sanitation businesses to be essential services and supported (i.e., through loan guarantees) to help financially weather reduced traffic flows. Train staff, and ensure business operations pay dividends once traffic resumes.
Project funding (and scope of work) focuses on building facilities rather than the operational details and technologies that encourage monitoring and financial sustainability.	Provide operator training and maintenance and management training. Support accelerating and integrating technologies for remote monitoring, automated cleaning, and digital payments. These can improve operating efficiencies and significantly enhance customer and employee offerings alongside traditional cash payments.
Disconnection between sanitation facility management and community leaders results in vandalism, suspicion, and conflict.	Sanitation businesses proactively communicate and work with community leaders. Leaders are trusted communication channels with the community, especially in times of crisis. If development organizations have previously run behavior change campaigns in the area, coordination building from their work is advisable.
Sanitation work is regarded as low status and only interests people with few alternatives and limited education or motivation. This can result in high staff turnover and loss of training and experience.	Elevate profile of sanitation professionals through living wages and attractive benefits. Provide professional training and resources—e.g., uniforms, personal protective equipment, and digital technology—to increase motivation.[a]
Staff are not provided with safe working conditions.	Include core labor standards as a contracting requirement. Encourage healthcare and/or insurance for workers.

[a] An example is the Asian Development Bank-assisted Tamil Nadu Urban Flagship Investment Program, which provided training for male and female municipal frontline sanitation workers on social safety and legal safeguard provisions, social dignity, personal safety, workplace hygiene, and equipment use. The rights-based approach to the training empowered participants and instilled pride, with female participants reporting increased confidence.

Source: Toilet Board Coalition. 2021. *Impact of the COVID-19 Pandemic on the Sanitation Service Provision and Delivery in Informal Settlements in Asia.* Geneva.

Challenges to Scaling Up

The case studies provide insight into the challenges of scaling up service delivery. These include the following:

(i) **Lack of suitable land.** The availability of land (especially in poor, high-density, and high-value areas, such as near markets) for constructing water or sanitation infrastructure is critical. For sanitation, the land needs to accommodate utilities and locations convenient for and highly visible to customers. In some places, it is possible to relocate a property to make space for community facilities. Land agreements may need to be negotiated with private owners, but there is also a risk that they may lose interest or divest that property to a new owner.

(ii)   **Limited utility will to collaborate.** The willingness of water and sanitation (or sewerage) utilities and local governments to partner with informal SMEs and social enterprises is crucial. The case studies show that this is possible and practical, and there are utilities and local governments that welcome the intermediary approach. Including these intermediaries may require additional support and adaptive procurement processes, or even appointing dedicated staff or establishing a special unit to address settlements (section 4.4). The first investments are likely to take time and meet the most resistance, but once the models are shown to be effective, resistance often reduces and scaling up is possible. Conversely, poorly implemented or unaffordable first investments, pilots, or demonstrations will not increase will and interest in collaboration.

(iii) **Lack of access to capital.** A major constraint for intermediaries is sufficient capital for water and sanitation infrastructure. Large capital investment projects for water and sanitation need to allocate resources for nonconventional components (such as contracts with local enterprises to act as intermediaries) to extend services to informal settlements. Such investments may initially be through cofinanced grant funding while a model is tested but later, if successful, lead to borrowing. Indonesia's *Sanitasi Berbasis Masyarakat* (SANIMAS or Sanitation by Community) community sanitation approach demonstrates a government's willingness to borrow capital after a model has been tested using grant funds.[12]

(iv)   **Minimal capacity-building ability.** Capacity building through training alone is inadequate to sustain a water or sanitation system. Ongoing support and on-the-job backstopping are needed for effective capacity building. It may take 2 or more years to commission, hand over, and support the local operators—as shown by parent organizations Eau et Vie, 1001fountains, Bhumijo, and Aerosan. Economies of scale through multiple sites in one city or area increase the cost-effectiveness of such support arrangements.

(v)   **Limited long-term commitment.** Support for social enterprises by subsidized NGOs involved in water supply—such as Eau et Vie and 1001fontaines—demands a long-term commitment. The commitment can be undermined by the NGO's decision to leave a country or change its policies or management. Local social enterprises may experience similar challenges where there is little leadership succession planning or the focus of the business shifts.

Recommendations for Improving Development Partner Effectiveness

ADB, other DFIs, and bilateral development partners can enable service delivery to informal settlement communities in several ways:

(i)   **Supporting intermediary partnerships.** Financing institutions and development partners should encourage client utilities to seek opportunities to supply bulk water and

---

[12] SANIMAS is an approach to community sanitation that was piloted in 2004 and 2005 by the World Bank-WSP working with the Bremen Overseas Development Agency, an NGO. From 2006, SANIMAS was scaled-up with Government of Indonesia investment funds. This included loan funds starting in 2009 from ADB's Metropolitan Sanitation Management and Health Project. The SANIMAS approach was adopted as part of the government's *Program Nasional Pemberdayaan Masyarakat* (National Program for Community Empowerment).

connect sewerage systems to small and medium-scale businesses, social enterprises, community-based organizations (CBOs), and NGOs as service providers to meet Sustainable Development Goals (SDG) and utility objectives and serve informal settlements.

(ii) **Transferring intermediary models.** Development banks and partners can use their knowledge platforms to share regionally about intermediary models, encourage local adaptation, and support tailoring them to different situations or countries and demonstrating their scalability.

(iii) **Adapting procurement processes.** Financing institutions and development partners should amend procurement processes to enable nonconventional service providers and contractors to apply and compete for contracts and funding. For example, the ADB-administered Promoting Urban Climate Change Resilience in Selected Asian Cities Project, financed by the Urban Climate Change Resilience Trust Fund, invited only international NGOs, in partnership with local NGOs, to tender for the implementation of the community-led resilience planning component. Development banks and government partners can make their systems less demanding for local subcontracts and thereby include a wider range of service providers while maintaining transparency and accountability.

(iv) **Supporting safeguards.** Development banks and partners should promote processes that improve inclusivity and achieve SDG targets by serving informal settlements and the most marginalized groups within those settlements. For example, ADB's Kolkata Environmental Improvement Investment Program connected the sewerage system to informal settlements and ensured that poor and women-headed households without tenure security were connected through project financing of the toilet construction cost. Vulnerable households were identified and certified by ward councilors or elected representatives.

(v) **Adopting sustainability indicators.** Financing institutions and development partners should use design and monitoring framework indicators that focus on the use and sustainability of facilities by the intermediary rather than just construction.

(vi) **Investing in capacity development.** Development banks and partners should support informal service providers to build skills and sustain services. This could involve reserving places for SMEs, social enterprises, NGOs, and CBOs in capacity-building activities of internationally financed projects (even if they are not direct beneficiaries). It could also mean including them in development-partner-initiated study tours, webinars, and workshops on water and sanitation topics. The participation of women should be especially encouraged.

(vii) **Seeking intermediary input.** Financing institutions and development partners should seek advice from experienced intermediaries to tailor contracting approaches. For market-based contracting of development bank operations and client-led investment projects, this will help to shape terms of reference and contract requirements. Dialogue and understanding between DFIs and intermediaries on ways of working and contracting needs will enable more appropriate terms of reference, procurement, and contracts to be developed.

(viii) **Encouraging scale up.** Development banks and partners should encourage intermediaries to scale up; reduce dependency; and seek full recovery of local management, operation, and maintenance costs. The case studies show that these actions are possible even when serving the poorest households.

Stepping up sanitation. Intermediaries deliver high-quality services while adapting and innovating their services to use technology for lower cost, service efficiency, and resilience (photo by Aerosan).

6

Summary

T he challenges of providing water and sanitation services to unserved informal settlement residents remain formidable and, in many Asian and Pacific towns and cities, are likely to increase with continued urban growth.

Working with intermediaries of different types provides a route for formal utilities and local governments to provide services to underserved or excluded residents. In return, the newly served areas can help grow the customer base, increase revenue, meet service targets, and in some situations, reduce nonrevenue water.

While the appropriate types of intermediaries vary by context, their roles are broadly similar. The case studies suggest that intermediaries such as social enterprises and SMEs can be effective and complementary service providers by collaborating with formal water and sanitation utility services and local governments. They can sustainably serve the needs of informal settlement residents by accessing bulk water or sewerage services from utilities and large private providers and leveraging external grant funding. However, development financing institutions, such as banks and multilateral and bilateral agencies, do not yet widely recognize that leveraging intermediaries offer significant benefits.

The lessons from the case studies should encourage governments, water supply and sanitation utilities, and development financing institutions to incorporate intermediate service providers as valuable partners for large-scale investments in water and sanitation service delivery. These case studies provide important, previously undocumented, evidence of intermediaries' flexibility, resilience, and vital role in contributing to achieving the SDGs in towns and cities in Asia and the Pacific.

Appendixes

Appendix 1
Case Studies

Case Study: Bhumijo, Bangladesh

Established in 2016, Bhumijo is a social enterprise providing public toilets in Bangladesh.

Public sanitation. Inside a toilet block (photo by Bhumijo).

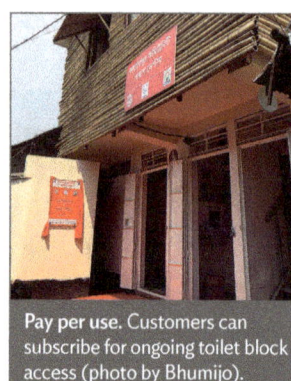

Pay per use. Customers can subscribe for ongoing toilet block access (photo by Bhumijo).

Institutional Arrangement	Bhumijo works with the private sector and government actors. Toilets are available in 28 locations.
Sanitation	Bhumijo is involved in planning, designing, researching, constructing, and maintaining public toilets. Its work includes renovating, constructing, and managing public and community toilets throughout Bangladesh. Bhumijo also aims to improve women's health by providing public access to quality sanitation services.
Operating Hours	Operating hours depend on where the toilet blocks are located. Some operate for 16 hours (6 a.m.–10 p.m.); others for 12 hours (10 a.m.–10 p.m.). In informal settlements, they operate 24 hours per day, 7 days per week.
Delivery Mode	The shared public sanitation hubs are connected to centralized sewerage where available. Where sewerage is unavailable, the hubs include septic tanks and on-site treatment. Waste goes to either the central sewerage or septic tanks.
Billing and Affordability	Governments, the private sector, or nongovernment organizations provide capital investments. Customers pay per use, but can subscribe weekly or monthly. Customers pay taka (Tk)5–Tk10 ($0.05–$0.10) per use or Tk50–Tk250 ($0.5–$2.5) per month for markets. Informal settlement users pay Tk100 ($1.0) per family per month. Average of 400 customers per day per facility.

Going digital. Payment using Bcash (photo by Bhumijo).

continued on next page

Case Study: Bhumijo, Bangladesh *continued*

Community and Coronavirus Disease (COVID-19) Pandemic Response	Bhumijo opened additional locations in Dhaka and increased users by 20%. They offered free use of facilities during lockdown periods. Bhumijo noted that a proactive community leader was key to successfully responding to the pandemic. In one community, COVID-19 protocols were more strictly observed and vaccinations accomplished faster than in the rest of the city because of the initiative the community leader showed.

Case Study: Aerosan, Nepal

Aerosan was established in Canada in 2014 and started operations in Nepal in 2018. Aerosan manages modern, women and disability-centric pay-for-use public toilets in crowded urban areas.

Institutional Arrangement	A common service provider–government relationship for community and public toilets is that service providers are given free lease of the land on which to operate the toilets. In Nepal, Aerosan experienced an acceleration of land concessions because of its increased focus and relevance. Nonetheless, Aerosan noted that the government should engage in performance-based management contracts with the private sector rather than lease the public toilets.
Sanitation	Aerosan public toilet hubs are modern, clean, safe, and women and disability-centric with anaerobic digesters that convert human waste to biogas. Each hub includes rainwater harvesting, gray water recycling, key sensors, wi-fi, sensor lighting, ventilation, non-touch taps and urinals, air hand drying, sanitary product dispensers, pad disposal, and baby changing facilities. Integral to the model are the anaerobic biogas digesters that convert toilet waste to biogas. The biogas plants are built in all locations, have low maintenance costs, and are constructed with local expertise and resources, making them sustainable, affordable, and feasible. The biogas is used for energy in the café or teahouse attached to the facility. After the fecal sludge has been composted, it is used as a nutrient-rich soil amendment for non-food plants in gardens in two toilet blocks.

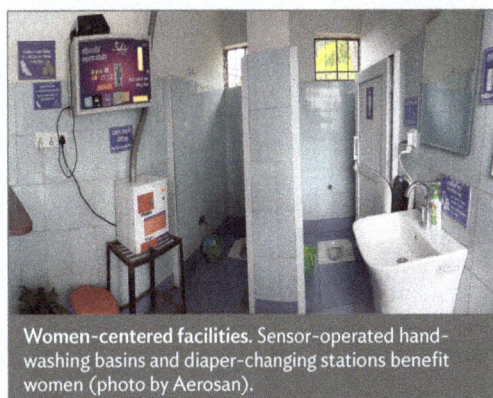

Women-centered facilities. Sensor-operated hand-washing basins and diaper-changing stations benefit women (photo by Aerosan).

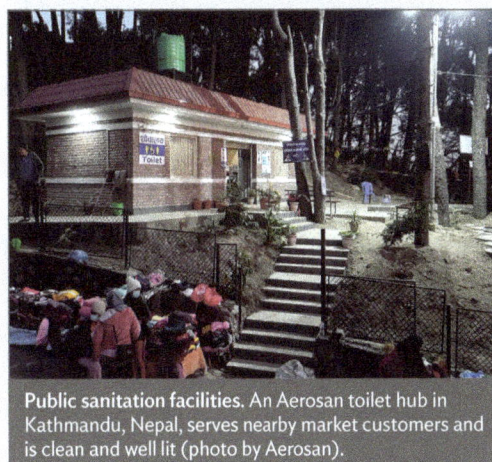

Public sanitation facilities. An Aerosan toilet hub in Kathmandu, Nepal, serves nearby market customers and is clean and well lit (photo by Aerosan).

continued on next page

Case Study: Aerosan, Nepal *continued*

Operating Hours	Normal operating hours are 6 a.m. to 9 p.m., 7 days a week.
Inclusive Designs and Policies	Free to people with disabilities, children, and older people. People who cannot pay are not required to.
Water Source	Rainwater and recycled water.
Delivery Mode	Aerosan facilities are shared sanitation hubs in six crowded urban locations, serving an average of 800 people daily.
Billing and Affordability	The capital costs are paid through grants and by the local municipality. The fees are set at a minimum and affordable rate, with different prices in different locations—normally Nepalese rupees (NRs)5 ($0.04) for urination and NRs10 ($0.04) for defecation. In Swoyambhu, the fee is NRs15 ($0.12) for both. Payments are made in cash or, since 2020, through a digital QR code system. However, most people prefer cash payments. Aerosan uses a pay-per-use model, but it is considering a subscription model for regular users, with 50% of the fees subsidized by Aerosan. Aerosan did not waive fees during the tightest COVID-19 pandemic lockdowns as its facilities were closed.
Community and COVID-19 Pandemic Response	Aerosan hires community members for the operation and maintenance of the facilities. The employment of local people helps decrease cases of theft, robbery, and misuse of facilities. Traffic through marketplace locations reduced to 1,200–1,300 users daily during the pandemic, substantially lower than the pre-pandemic average of 3,000–5,000 users.

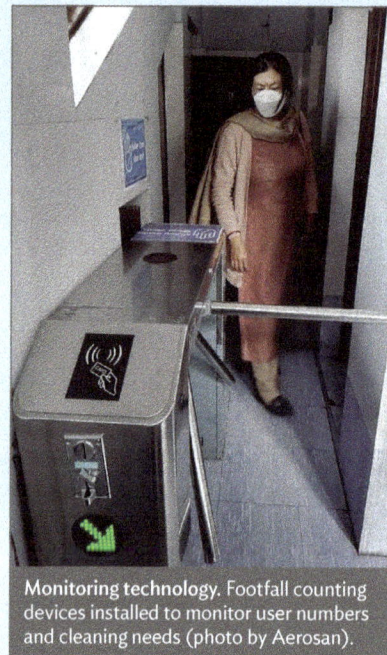

Monitoring technology. Footfall counting devices installed to monitor user numbers and cleaning needs (photo by Aerosan).

Case Study: Suvidha, Mumbai, India

Suvidha (meaning confidence) is a partnership offering community hygiene centers that provide access to quality sanitation services to urban residents living under the poverty line.

Institutional Arrangement	Suvidha is a public–private partnership of Hindustan Unilever Limited launched in 2016 in close consultation with the Municipal Corporation of Greater Mumbai. Suvidha centers are built in highly congested areas where there is maximum need. The municipal government provides land, and capital expenditure is by Hindustan Unilever Limited and HSBC. Nongovernment organizations and social enterprises run the centers with technical support from Hindustan Unilever Limited. Subscription packages are sold to residents, which cover operation and maintenance expenses.
Sanitation	The Suvidha model aims to transform urban community toilets into clean, hygienic, and self-sustainable centers that provide economic, social, and environmental benefits. In addition to providing an avenue for community engagement and behavior change programs, these facilities lead to high satisfaction and ownership among users.
Delivery Mode	Five Suvidha sanitation hubs collectively serve more than 120,000 people every year. The hubs are situated in crowded, high-density slum settlements where people do not have space for private facilities or cannot afford them. Each Suvidha center uses piped and rainwater sources. They save 4 million liters of water annually using a closed-loop recycling of water from the handwashing and laundry facilities. Water from handwashing and washing machines is treated and reused for the toilets. The Suvidha center in Ghatkopar, opened in 2021, treats and reuses gray and black water. It is designed for 20,000 users and saves more than 10 million liters of fresh water annually by reusing treated wastewater. It uses a hybrid treatment system based on the decentralized wastewater treatment system approach.
Inclusive Designs and Policies	Inclusively designed for use by children, older people, and people with disabilities. The model allows older people, children, and people with disabilities to use the service for free.

Toilets for people with disabilities. Ramps, handrails, and low sinks help more people to access quality toilets (photo by Suvidha).

continued on next page

Case Study: Suvidha, Mumbai, India *continued*

Billing and Affordability	The Suvidha centers become operationally self-sustainable within 6 months of inception. At each location, about 15 to 20 families who cannot afford the service charges are identified in advance and provided access to the center free of charge.

Service	Public Toilets / Service Rates	Suvidha Rates	Remarks (₹1 = $0.012)
Toilet service monthly pass	₹100 ($1.20)	₹150 ($1.83)	Family of 5 to 6
Toilet service pay and use	₹5 ($0.06)	₹3 ($0.04)	
Shower service	₹10 ($0.12)	₹10 ($0.12)	5–7 minute shower
Drinking water per liter	₹20 ($0.24)	₹1 ($0.01)	
Laundry service	₹840 ($10.28)	₹55 ($0.67)	7 kilograms of clothes (1 wash load)

₹ = Indian rupee, $ = United States dollar.

COVID-19 Pandemic Response	Suvidha saw a 150% increase in the number of community toilet users in informal settlements during the COIVD-19 pandemic as more individuals were at home and prioritized using a higher standard of hygienic toilets. During the COVID-19 lockdown in Mumbai, the Suvidha toilet service was declared free for the community for 6 months.

Case Study: Laguna Water, Laguna Province, Philippines

Established in 2009, Laguna Water is a water supply, used water, and environmental services provider in the Philippines.

Institutional Arrangement	Laguna Water is a joint venture between the Provincial Government of Laguna and Manila Water Philippine Ventures. The company operates in nine cities and municipalities in the Philippines and has 100,000 water supply connections. Laguna Water can treat 11 million liters of sewage daily and 70 cubic meters of septage daily.
Sanitation	In 2018, Laguna Water launched its sanitation program, *Tamang Sanitasyon Equals Kalusugan, Kalinisan, at Kaunlaran ng Bayan*. The program uses a three-pronged approach to provide sanitation services to people of all socioeconomic groups. Services include sewage management, desludging or septic tank cleaning services, and communal toilets or portable toilets. 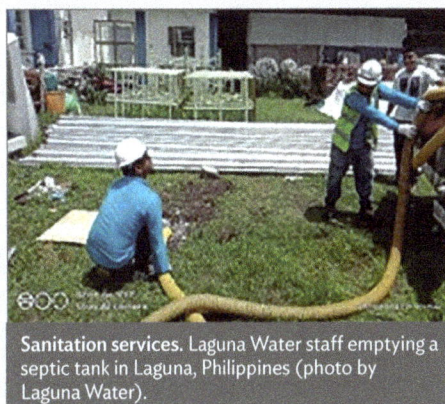 **Sanitation services.** Laguna Water staff emptying a septic tank in Laguna, Philippines (photo by Laguna Water).
Delivery Mode	Laguna Water provides sewerage services and fecal sludge management of septic tanks where there is no sewerage connection (or a connection is impossible).
Billing and Affordability	All water supply customers have levied an "environmental charge" to cover sanitation services, irrespective of whether sewers are in their neighborhood. Laguna Water takes full responsibility for the wastewater generated from the water they supply. Charges for sanitation are based on the volume of water used. The basic volumetric water charge (excluding value-added tax or other taxes) is levied as an environmental charge of 20% for desludging services and 50% for sewered areas. The model emphasizes environmental protection and promotes that everyone should contribute according to their means. Everyone pays something, despite the subsidy for poor households and discounts for older people. Discounts are shown on the water bill. The move to online payments and the reduction in cash is evident in Laguna Water's accounts, with a decline in cash payments since April 2019 balanced by the increase in online payments through a local app called GCash.
Community Engagement	Laguna Water participates in community programs such as *Tubig Para sa Barangay* (Water for the Poor); *Lingap Eskwela* (Water and Sanitation for Public Schools); *Kabuhayan Para sa Barangay* (Livelihood Program); Information, Education, and Communication Campaign on Used Water; and *Tamang Sanitasyon Equals Kalusugan TSEK* (Sanitation for Marginalized Communities).
COVID-19 Pandemic Response	Laguna Water stopped disconnections and deferred late payment fees to its customers in compliance with the Bayanihan Act. During strict lockdowns, the act mandated a grace period and a 3-month installment for water bills. Laguna Water also provided a grace period during less-strict lockdowns.

Case Study: Shobar Jonno Pani Ltd in Chattogram, Bangladesh

Shobar Jonno Pani (SJP)—or "Water for All" in Bengali—is a local social enterprise founded in 2010 and responsible for (i) developing the water infrastructure network and its maintenance (piped drinking water with metered connections on household premises), (ii) building toilets and drainage for improved sanitation (either within a sanitary block or as individual latrines), (iii) providing waste management services, (iv) ensuring water quality, and (v) billing and collecting the payments for water and latrine services. SJP began supplying water to Bhashantek, a low-income community of Dhaka, in 2013 and extended to the No. 9 Community in Chattogram in 2018. Piped water is provided to more than 2,200 households (10,000 people) in Bangladesh.

Institutional Arrangement	A mini utility is a delegated management arrangement between the utility and a local service operator—SJP. A memorandum of agreement has been in place for 10 years between SJP and the utility Chattogram Water and Sewerage Authority. Such long-term commitment allows social enterprises to recover operational costs and cover ongoing technical assistance. The French nongovernment organization (NGO) Eau et Vie provides technical support, coordination, and advocates for governments, water regulators, municipalities, and leaders to deploy the model in additional low-income communities. A local registered NGO, Water and Life Bangladesh, delivers complementary modules and is accountable for hygiene promotion, firefighting, and community empowerment activities. It also conducts studies to assess the local needs and gather feedback from the beneficiaries.
Water Supply and Quality	Water is supplied from a legal connection to the municipal network operated by Chattogram Water and Sewerage Authority. Through four bulk meters and a connected pipe network, SJP commits to providing a sufficient quantity of at least 50 liters of water per person per day with 24/7 availability, meeting national water quality standards. Bulk water receives additional chlorination with a dosing pump (at a pumping station after the bulk meter). The two internal water networks (necessary because of steep hills and a bisecting railway line) were designed by an Eau et Vie engineer and built to optimize pressure and ensure quality while considering technical constraints. 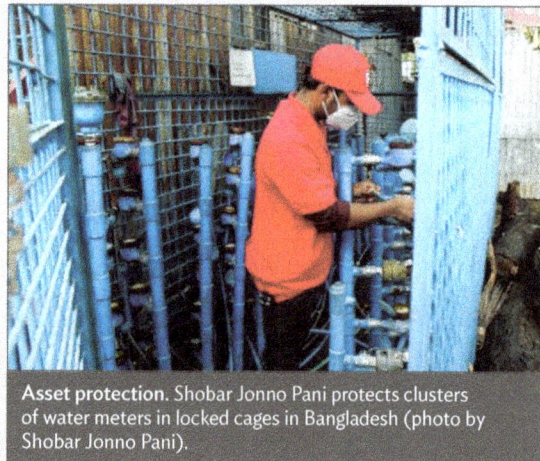 **Asset protection.** Shobar Jonno Pani protects clusters of water meters in locked cages in Bangladesh (photo by Shobar Jonno Pani).
Delivery Mode	Households get an individual metered water and tap connected to the piped network. The high-quality precision meters are placed in clustered meter cages to prevent damage and vandalism.

continued on next page

Case Study: Shobar Jonno Pani Ltd in Chattogram, Bangladesh *continued*

Billing and Affordability	The water service is billed weekly according to consumption, with a tariff comprising the utility's water price and SJP's operation and maintenance costs. The price per cubic meter is adapted to the average household income in the area. A common billing system bundles the water bill with solid waste and sanitation services. Fees are collected at home at a frequency appropriate to household incomes, making payments easy for customers. The rate is affordable for low-income residents, particularly as payments are made in small regular installments. The fee collection is 98%. In cases of nonpayment, the process is strict, with a disconnection notice once a certain amount of arrears is reached and disconnection from the water network 2 weeks later. SJP's extreme poor policy provides a waiver for the access fee and 50 liters per person per day for free for up to four people.
	Connection fees are staggered via a microcredit, and the frequency and duration of repayment depend on income. Connections are made within 2 weeks of signing the contract.
Community Engagement	The focus on quality compliance and service delivery produces loyal customers who trust the social enterprise. The reliable water source also improves emergency and firefighting responses, especially in densely populated informal settlements where fires prove devastating.

Firefighting and women's empowerment. With access to a reliable water supply, women in Bangladesh learn how to operate firefighting equipment and hoses—critical to emergency response in dense poor settlements (photo by Shobar Jonno Pani).

With water provided by SJP, Water and Life delivers firefighting equipment and training to volunteers, with support from local fire authorities. Volunteer brigades are then able to respond in case of fire.

COVID-19 Pandemic Response	SJP continued to supply water during the COVID-19 pandemic due to an uninterrupted supply from Chattogram Water and Sewerage Authority. A water relief program provided 7.5 cubic meters of free water to households in June–July 2020 and August 2021 to support quality water consumption. Emergency repair services for leakages or other technical issues were provided by staff living in the settlement. A 24/7 hotline was set up for the community to report any issues, and plumbers were provided with 2 months' supply of chlorine and water quality tests.

Case Study: Tubig Pag-Asa, Inc., Guizo Community, Mandaue, Philippines

Tubig Pag-Asa, Inc. (TPA), or "Water of Hope" in Filipino, is a social business that aims to deliver clean and safe drinking water to urban populations living below the poverty line. TPA has been working in the Philippines since 2009, servicing 4,855 connections or 6,058 households in 18 communities. The Guizo Community in Mandaue City was connected in 2018, with 331 connections serving 1,522 people.

Institutional Arrangement	TPA partners with local operator Metro Cebu Water District (MCWD). The arrangement has been formalized through a 10-year memorandum of agreement (MOA) between TPA, MCWD, and the local authority. The MOA states the responsibilities of each party, the cost of the water connection for the customers, and the permits and requirements requested for TPA to operate. The MOA allows MCWD to adopt TPA's network infrastructure 10 years after the bulk water supply has been installed, subject to the community's ability and willingness to pay their monthly water bills. TPA works in partnership with nonprofit organization Water and Life Philippines, which provides complementary interventions in the communities, such as hygiene awareness, firefighting and emergency management, solid waste management, and community empowerment.
Water Supply and Quality	Treated water is supplied to the community by MCWD to a bulk meter at the entrance to the community. TPA may provide additional treatment through chlorination if regular water quality tests indicate this is necessary to comply with national water quality standards. Most households do not use TPA water for drinking and instead buy purified water. The internal water network was designed by an engineer and built to optimize pressure and ensure quality while considering technical constraints. 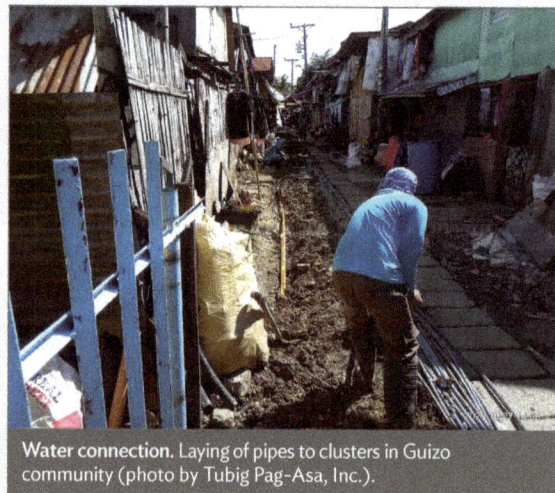 Water connection. Laying of pipes to clusters in Guizo community (photo by Tubig Pag-Asa, Inc.).
Delivery Mode	Individual household connections consist of installing individual meters (located in collective and secure clusters), burying individual pipes (from the meter to the household), and possibly cementing the alleys after burial to protect pipes from leaks.

continued on next page

Case Study: Tubig Pag-Asa, Inc., Guizo Community, Mandaue, Philippines *continued*

Billing and Affordability	TPA charges a staggered connection fee, with repayment frequency determined by income. Water tariffs are competitive, adapted to the area's average household income, and include the bulk water charge plus TPA overheads (Philippine peso [₱]1 [$0.02] for every 20 liters). Meter readings are conducted daily. Water bill collection is done door-to-door, on a daily or weekly basis to suit customers. A mobile application has been developed to help with bill collection and reporting. TPA operates a savings scheme whereby the customer can pay more than the daily bill, with the savings visible on the receipt and smartphone account. The scheme has helped move some customers from daily to weekly payments.	 **Rising above.** Tubig Pag-Asa, Inc. has overcome Guizo community's dense living conditions and technical challenges to provide a quality water supply service (photo by Tubig Pag-Asa, Inc.).
Community Engagement	The approach aims to empower users and prepare them for future management by MCWD. With water provided by TPA, Water and Life provides community engagement and consultation, hygiene and menstrual hygiene awareness, and firefighting training. TPA provides local work opportunities for community members.	
COVID-19 Pandemic Response	TPA provided a continuous water supply during the COVID-19 pandemic. A water relief program provided 7.5 cubic meters of free water to households from June to July 2020 to support quality water consumption. An extreme poor policy was introduced permanently.	

Case Study: Motu Koita Assembly, Pari Village, Port Moresby, Papua New Guinea *(under development)*

The Motu Koita Assembly (MKA) is the representative authority for the Motu and Koitabu people, the traditional landowners of the greater Port Moresby area. The Moita and Koitabu people live in nine urban villages, mainly relying on expensive carted water, self-supply, or unimproved sources. Pari village is a demonstration village for the delivery of water supply services that can be used to establish replicable and scalable models for other traditional villages and the rapidly growing informal urban space in major towns and cities of Papua New Guinea (PNG).

The project is awaiting the installation and testing of a 2.8-kilometer pipeline leading to Pari village. This will be installed and tested by the utility Water PNG in 2023. An interim solution is to truck water to the kiosks, where it will be sold by kiosk operators, who will take over selling piped water when the pipeline is operational.

Institutional Arrangement	MKA is a legal entity established by an act of parliament. MKA has a memorandum of agreement with the utility Water PNG, which assigns responsibility for operations, maintenance, billing, and fee collection to the MKA villages, in exchange for a minimum bulk water service provision to the villages at a discounted rate for 3 years. Kiosk operators are contracted to Pari Water Services, which is a business entity of MKA. A full-time manager and accountant will oversee the water system and billing. WaterAid provides technical support to improve water distribution, financial management, and governance arrangements.
Water Supply and Quality	Water PNG will supply bulk-treated water at 40 liters per person per day to national water quality standards.
Delivery Mode	Water will be delivered through a piped network to 9,000-liter overhead tanks at seven disability-accessible water kiosks. Household connections will be possible where previous connections existed. The utility's bulk meter is at the community's boundary, with meters at the overhead tanks and on taps at the retail point. Meters are in a locked cage at the kiosk.

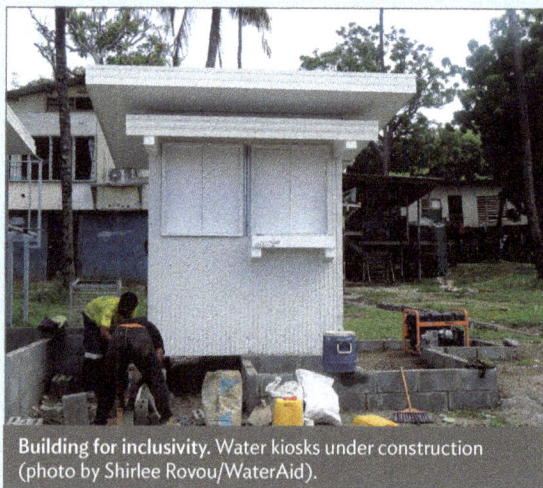

Building for inclusivity. Water kiosks under construction (photo by Shirlee Rovou/WaterAid).

Billing and Affordability	The water tariffs for kiosks and piped water are agreed upon with the utility and include the wholesale bulk water rate plus the cost of operation and maintenance and some replacement costs for the water supply system. While the tariff is 10 times more expensive than a formal metered Water PNG connection, it is affordable for Pari residents as it is one-tenth of the rates informal water sellers charge. Electronic payment methods allow customers to pay their bills at the kiosk anytime.

continued on next page

Case Study: Motu Koita Assembly, Pari Village, Port Moresby, Papua New Guinea *continued*

Community Engagement	Water will be supplied to the Pari community and neighboring informal settlement. Community engagement is conducted through surveys; meetings; and a water, sanitation, and hygiene committee. The water system also provides local employment.
COVID-19 Pandemic Response	Pari village's water supply was not operational during the COVID-19 pandemic. MKA and WaterAid arranged for water to be trucked into the community for handwashing at schools and public points and undertook awareness-raising activities such as screening behavior change videos and holding meetings.

Case Study: 1001fontaines, Cambodia

The 1001fontaines Cambodia initiative was launched in 2005 to improve the health of poor people living in remote areas of the country. Serving 260 sites in Cambodia, the operational model has been adapted to peri-urban and urban settings in Madagascar and Viet Nam, bringing the total sites to 288. 1001fontaines provides water kiosks as a complementary service to a piped water supply.

Institutional Arrangement	The approach is a public–private–nonprofit partnership. Water kiosks are set up in underserved areas selected by the local community and authorities. Local entrepreneurs manage the production and distribution of safe water and are trained in water purification, marketing, finance, entrepreneurship, and team management. Through a franchising model, the water kiosks receive ongoing technical and business support from regional and national teams. 1001fontaines has a global team supporting local entities with strategic guidelines, knowledge transfer, advocacy, and partnership development. 1001fontaines collaborates with local governments through commune authorities.
Water Supply and Quality	The model enables the decentralized production of safe drinking water in communities through sustainable water kiosks. Raw water is extracted from local sources such as surface water ponds, rivers, or dug wells and treated to World Health Organization standards at a small facility on public land. Safe, convenient water. Water is treated and bottled at kiosks and delivered to customers' homes to ensure accessibility for the last mile (photo by 1001fontaines).
Delivery Mode	The treated water is bottled into reusable 20-liter containers and delivered to consumers' homes.
Billing and Affordability	The 20-liter water bottles are sold at an affordable price ($0.02 per liter) as recommended by the World Health Organization. Revenues cover water kiosk operating costs and the entrepreneur's salary, ensuring service sustainability.
COVID-19 Pandemic Response	1001fontaines worked with commune authorities to identify households most in need and provided them with free 20-liter bottles, hygienic products, and safe water refills for 3 months through kiosk entrepreneurs.

Appendix 2
Additional Notes on Methodology

Toilet Board Coalition Case Studies

The Toilet Board Coalition's case studies were selected if they met four criteria: the service provider was operational before the coronavirus disease (COVID-19) pandemic (prior to March 2020); facilities or services are provided in informal settlements or low-income communities; a unique business model or sanitation solution is offered; and service providers are willing to participate in the study. From the five selected service providers, study participants included business leaders (owners, managers, or representatives), frontline workers (staff who clean and maintain toilet facilities and/or deal directly with sanitation facility users), and customers (people who use the sanitation facilities or services). The five business leaders (one per enterprise) completed a brief survey, participated in a semi-structured interview with the Toilet Board Coalition, and joined a focus group discussion. Frontline workers—five female and 10 male employees—completed a short survey. A total of 160 female and 148 male customers, selected using simple random sampling, completed a short survey on sanitation use and priorities.

Eau et Vie Case Studies

Eau et Vie selected their case studies based on shared characteristics: (i) the same number of beneficiaries, (ii) connected to the water network in 2018, and (iii) the same period of intervention by Eau et Vie. Data-collection methods included women-only and community-level focus group discussions (31 women and 9 men participated in Bangladesh and 21 women and 11 men in the Philippines) and key informant interviews with four government stakeholders and two water delivery providers.

References

1001fontaines. 2020. *Cambodia Economic Sustainability Report*. Paris.

_____. n.d. *Water Kiosk Delivery Service: Bridging the Safe Drinking Water Gap in Underserved Areas*. Paris.

AECOM International Development, Inc. and the Department of Water and Sanitation in Developing Countries (Sandec) at the Swiss Federal Institute of Aquatic Science and Technology (Eawag). 2010. *A Rapid Assessment of Septage Management in Asia: Policies and Practices in India, Indonesia, Malaysia, the Philippines, Sri Lanka, Thailand, and Vietnam*. Washington, DC: United States Agency for International Development.

Ahlers, R. et al. 2014. Informal Space in the Urban Waterscape: Disaggregation and Co-production of Water Services. *Water Alternatives*. 7 (1). pp. 1–14.

Asian Development Bank (ADB). 2018. *Strategy 2030: Achieving a Prosperous, Inclusive, Resilient, and Sustainable Asia and the Pacific*. Manila.

_____. 2019. *Asian Development Outlook 2019 Update: Fostering Growth and Inclusion in Asia's Cities*. Manila.

_____. 2022. *Strategy 2030 Water Sector Directional Guide—A Water-Secure and Resilient Asia and the Pacific*. Manila.

Cambodia News English. 2021. Siem Reap Residents Illegally Connecting Sewers to New Network. *Cambodia News English*. 17 March. https://cne.wtf/2021/03/17/siem-reap-residents-illegally-connecting-sewers-to-new-network/.

Chen, M.A. and V.A. Beard. 2018. *Including the Excluded: Supporting Informal Workers for More Equal and Productive Cities in the Global South*. Working Paper. Washington, DC: World Resources Institute.

Conan, H. 2004. Small Piped Water Networks: Helping Local Entrepreneurs to Invest. *ADB Water for Life Series*. No. 13. Manila: ADB.

DOHWA Engineering Co. Ltd. 2021. *Dhaka Water Supply Network Improvement Project (DWSNIP) Consultancy for Project Development Facility (PDF)* Contract No. DWSNIP / PDF / 03.5. Feasibility Report.

Frauendorfer, R. and R. Liemberger. 2010. *The Issues and Challenges of Reducing Non-Revenue Water.* Manila: ADB.

Garrick, D. et al. 2019. *Informal Water Markets in an Urbanising World: Some Unanswered Questions.* Washington, DC: World Bank.

IGI Global. 2021. *What is an Informal Service Provider?* https://www.igi-global.com/dictionary/ existing-realities-and-sustainable-pathways-for-solid-waste-management-in-ghana/78894.

Jayaramu, K.P., A. Devadiga, and B.M. Kumar. 2015. Connecting the Last Mile: Water Access Policy in Action. *Waterlines.* 34 (2). pp. 156–173.

Lardoux de Pazzis, A. and A. Muret. 2021. *The Role of Intermediaries to Facilitate Water-Related Investment.* OECD Environment Working Papers. No. 180. Paris: OECD Publishing.

McDonald D.A. and G. Ruiters. 2007. Rethinking Privatisation: Towards a Critical Theoretical Perspective. *Beyond the Market: The Future of Public Services.* Amsterdam: Transnational Institute.

McIntosh, A.C. 2003. *Asian Water Supplies: Reaching the Urban Poor.* ADB and International Water Association.

Mitlin, D. and A. Walnycki. 2020. Informality as Experimentation: Water Utilities' Strategies for Cost Recovery and their Consequences for Universal Access. *The Journal of Development Studies.* 56 (2). pp. 259–277.

Mitlin, D. et al. 2019. *Unaffordable and Undrinkable: Rethinking Urban Water Access in the Global South.* Working Paper. Washington, DC: World Resources Institute.

Mukhija, V. 2016. *Rehousing Mumbai: Formalizing Slum Land Markets through Redevelopment.* Philadelphia: Penn Institute for Urban Research.

Nakamura, S. 2014. Impact of Slum Formalization on Self-Help Housing Construction: A Case of Slum Notification in India. *Urban Studies.* 51 (16). pp. 3420–3444.

Office of the High Commissioner for Human Rights. 2020. COVID-19 Pandemic and the Human Rights to Water and Sanitation. *UN-Water.* 17 November. https://www.unwater.org/covid-19-pandemic-and-the-human-rights-to-water-and-sanitation/.

Peal, A. and S. Drabble. 2015. *Stand-alone unit or mainstreamed responsibility: how can water utilities serve low-income communities?* London: Water & Sanitation for the Urban Poor.

Schrecongost A. and K. Wong. 2015. *Unsettled: Water and Sanitation in Urban Communities of the Pacific.* Washington, DC: World Bank.

Shagun. 2019. Manual Scavenging has Gone Underground in India: WHO. *Down to Earth.* 14 November. https://www.downtoearth.org.in/news/waste/manual-scavenging-has-gone-underground-in-india-who-67751.

Singh, S. 2019. *Achieving Sustainable Sanitation in Asia*. Policy Brief. No. 2019-4. Tokyo: Asian Development Bank Institute.

Sinharoy, S., R. Pittluck, and T. Clasen. 2019. Review of Drivers and Barriers of Water and Sanitation Policies for Urban Informal Settlements in Low-Income and Middle-Income Countries. *Utilities Policy*. 60.

Toilet Board Coalition (TBC). 2021. *Impact of the COVID-19 Pandemic on the Sanitation Service Provision and Delivery in Informal Settlements in Asia*. Geneva.

United Nations Human Settlements Programme (UN-Habitat). Urban Indicators Database. https://data.unhabitat.org/pages/housing-slums-and-informal-settlements (accessed 7 August 2023).

_____. 2015. Informal Settlements. *Habitat III Issue Papers* 22.

UNICEF and WHO. 2018. *Core Questions on Drinking Water, Sanitation and Hygiene for Household Surveys: 2018 Update*. New York: United Nations Children's Fund (UNICEF) and World Health Organization.

United Nations General Assembly. 2010. *Human Right to Water and Sanitation*. Resolution 64/292.

WaterAid. 2009. *Water utilities that work for poor people: Increasing viability through pro-poor service delivery*.

_____. 2016. *Water: At What Cost? The State of the World's Water 2016*.

World Bank, International Labour Organization, WaterAid, and World Health Organization. 2019. *Health, Safety and Dignity of Sanitation Workers: An Initial Assessment*. Washington, DC: World Bank.

World Bank Group and Water and Sanitation Program. 2015. *Improving Onsite Sanitation and Connections to Sewers in Southeast Asia: Insights from Indonesia and Vietnam*. Washington, DC: World Bank.

_____. 2013. *Getting to Scale in Urban Water Supply:* Topic Brief. London.

www.ingramcontent.com/pod-product-compliance
Lightning Source LLC
Chambersburg PA
CBHW050051220326
41599CB00045B/7373